ADVANCE PRAISE FOR *Pedagogy for Restoration*

"David Krzesni's *Pedagogy for Restoration* is a deeply personal work that raises important questions about the relationship between social and ecological degradation. The book is incredibly timely as such questions and examinations are proving to be among the most trenchant of our time. Krzesni's focus on the politics of hope, empathy, and moral education not only provides a unique frame through which to consider the structures of oppression and liberation but also offers students of education and social change what they have long asked for—a pedagogy of restoration."

—*Sandy Grande, Associate Professor, Co-Chair, Education Department, Director, Center for the Comparative Study of Race and Ethnicity, Connecticut College*

"At a time where much looks bleak for people and their relationships with themselves, others, and with the earth, we need narratives of renewal, purpose, and possibility. We need guides to remind ourselves of our own potential to heal broken relationships. In *Pedagogy for Restoration*, David Krzesni is such a guide. This book will challenge your identity as an educator and provide you with pathways for your own revitalization in connection with others."

—*David A. Greenwood, Canada Research Chair in Environmental Education, Director, Centre for Place and Sustainability Studies, Faculty of Education, Lakehead University*

"David Krzesni's *Pedagogy for Restoration* is a well-researched piece of literature.... He suggests that education can be a powerful tool for effecting structural change, remaining deeply sensitive to mechanisms of oppression and discrimination, the power of language, and the need for a hope-based, informed approach to social change.... Krzesni cites research on human qualities such as empathy and altruism to highlight people's innate propensity for caring and responsibility. He believes in the human capacity for change, and believes that, through structured guidance and governance, positive qualities can be emphasized and people can be encouraged to act on these characteristics. While acknowledging the limitation to human love, empathy, and prosocial behavior, Krzesni argues that, with the right social conditions and with practices of nurturing and empowerment (especially in formative years), people's capacity to carry out a 'pedagogy for restoration' is possible."

—*Samuel P. Oliner, Co-author (with Pearl Oliner) of* The Altruistic Personality *and* Toward a Caring Society; *Author of* Altruism, Intergroup Apology, Forgiveness, and Reconciliation *and* The Nature of Good and Evil

"In this book by David Krzesni, a powerful new moral voice addresses the intersection of environmental education, social justice, and human rights. Boldly affirming both critical social theory and human moral potentials to overcome every exclusion, this bracing and exacting fugue freshly points out melodies of deepest humanity, hearing voices that unite ecology and people in powerful harmonies. It should be required/inspired reading (listening, singing) for the next generation of critical socio-environmental change agents."

—*Gene Myers, Professor, Environmental Studies, Huxley College, Western Washington University*

Pedagogy *for* Restoration

COUNTERPOINTS

Studies in the Postmodern Theory of Education

Shirley R. Steinberg
General Editor

Vol. 503

The Counterpoints series is part of the Peter Lang Education list.
Every volume is peer reviewed and meets
the highest quality standards for content and production.

PETER LANG
New York • Bern • Frankfurt • Berlin
Brussels • Vienna • Oxford • Warsaw

DAVID KRZESNI

Pedagogy for Restoration

Addressing Social and Ecological Degradation through Education

PETER LANG
New York • Bern • Frankfurt • Berlin
Brussels • Vienna • Oxford • Warsaw

Library of Congress Cataloging-in-Publication Data

Krzesni, David, author.
Pedagogy for restoration: addressing social and ecological degradation through education / David Krzesni.
pages cm. — (Counterpoints: studies in the postmodern theory of education; vol. 503)
Includes bibliographical references and index.
1. Environmental education—Social aspects. 2. Environmental degradation—Study and teaching. 3. Social justice—Study and teaching. I. Title.
GE70.K79 363.70071—dc23 2015017298
ISBN 978-1-4331-3155-4 (hardcover)
ISBN 978-1-4331-3154-7 (paperback)
ISBN 978-1-4539-1651-3 (e-book)
ISSN 1058-1634

Bibliographic information published by **Die Deutsche Nationalbibliothek**.
Die Deutsche Nationalbibliothek lists this publication in the "Deutsche Nationalbibliografie"; detailed bibliographic data are available on the Internet at http://dnb.d-nb.de/.

Cover image by David Krzesni

29 Broadway, 18th floor, New York, NY 10006
www.peterlang.com

To my Oma, Rosalie Roozendaal Krzesni (1913–2004).
It is significantly in her memory that I have been motivated both in this project and in the struggle for a better world.

CONTENTS

PREFACE

> The United States is a nation defined by its original sin: the genocide of American Indians. Everything afterward is just another chapter in the fall from grace.... No reparation, no penance, no atonement can ever erase the eternity of genocide. Life ever after will be forever stained by the attainment of this "carnal knowledge." Such an inauspicious beginning raises significant questions about the viability of this so-called democratic experiment: Is it possible for democracy to grow from the seeds of tyranny? Can the "good life" be built upon the death of thousands? (Grande, 2004, p. 31)

I began this project as a student at Western Washington University, an institution built on occupied territory of the Lummi Nation. As a graduate of that institution, I find myself to be complicit in the continuing process of colonization of America. But I refuse to silently collude by ignoring inconvenient and uncomfortable truths. Colonization is not a historical event and the oppressive, violent, tragic, and painful results of the ongoing invasion of America are plain to see for anyone who dares look with a critical eye. They are also easily ignored. Many of us have the luxury to overlook or even deny that people are suffering and dying at the expense of our own privilege and ignorance. To begin this project of restoration in any other way than to acknowledge colonization would be not only fraudulent, but also demeaning of those who suffer this injustice, to those who are targeted by oppression, and to all of those already working for restoration.

I have constantly wondered whether any vision for a better world is inevitably short sighted when imagined within the academy. As one of the great thinkers of our time, Albert Einstein was not only concerned with theoretical physics, but as a German Jewish refugee, was also deeply concerned with civil rights. Having been alive during the atrocities of World War II, including the United States' use of the atomic bomb on Hiroshima, Einstein worried that history would repeat itself. Einstein (1946) wrote:

> Today we must abandon competition and secure cooperation. This must be the central fact in all our considerations of international affairs; otherwise we face certain disaster. Past thinking and methods did not prevent world wars. Future thinking *must* prevent wars.

The "past thinking" that Einstein denounced was the dominant Western way of thinking often referred to as Western Modern Science[1] (WMS) (Ogawa, 1995). Einstein did not denounce the scientific and technologic outcomes of WMS that made the invention of the atomic bomb possible, but the competition and domination that may be inherent in modernism. Warfare was/is much less common in hunter-gatherer societies and warfare of the scale and magnitude that Western society has experienced likely arose with agriculture and industrialization (Abrams, Coast Community College District, & KOCE-TV, 2002).

This project of restoration is centered in academia and conducted in a manner congruent with WMS. This work may be inseparably associated with systems of oppression structured to produce and communicate knowledge in ways that maintain the status quo (Horkheimer & Adorno, 1944/2002). Furthermore, it has been said that "the master's tools will never dismantle the master's house" (Lorde, 1984, p. 110). As a Black[2] lesbian feminist, Lorde (1984) was writing about the hierarchy, oppression, and relations of power within (and beyond) feminism. Lorde suggested that the presence of such "tools of a racist patriarchy" would undermine feminist efforts, ensuring "that only the most narrow parameters of change are possible and allowable" (Lorde, 1984, p. 110).

Many interpret Lorde's (1984) essay more generally and suggest that academia itself is one of the master's tools. I recognize the presence of hierarchy, power, and oppression in academia, but that doesn't condemn the entire institution. I think Lorde was referring to the social relations associated with institutions such as academia rather than those institutions in their entirety. We must be vigilant and critical of the structures we operate within and careful that we don't use the master's tools in ways that support the master's cause (oppression). If we challenge the master's rule and constantly work to refuse to conduct ourselves according to the master's rules, I believe we can fight oppression within the academy. In fact, academia is one of many fronts where we must work to bring about structural change. Thus, I carefully and critically proceed within the WMS framework while attempting to leave space for multiple epistemologies.

However, it is problematic to presume that one could envision any utopian future from a space in which these problems exist and deeply shape our minds. Even if this could be done sufficiently to imagine an incremental improvement, the "timeless nature of the gap between the world of our aspirations, hopes, and dreams and the world we create with our policies, practices, and every day actions" (Loeber, van Mierlo, Grin, & Leeuwis, 2007) must be addressed and somehow bridged. Glasser (2007) offered a poem written in Egypt four thousand years ago illustrating the timeless nature of this gap.

> To whom can I speak today?
> The gentle man[3] [*sic*] has perished
> The violent man [*sic*] has access to everybody.
> To whom can I speak today?
> The iniquity that smites the land
> It has no end.
> To whom can I speak today?
> There are no righteous men [*sic*]
> The earth is surrendered to criminals. (p. 40)

The tragedy is not in the identification of iniquity, but in the loss of hope. The quote clearly identifies the crisis as a moral one with physical manifestations realized when humans collectively fail to live up to their potential. It also illustrates that much of our modern concerns over the future have plagued generations of humans and are not contemporary issues. Multiple religions and worldviews hold that there was a time in history when our potential was more fully realized. Much of the philosophy and religion of modern Western civilization is consumed with understanding how, when, and why we went wrong (e.g., the Fall in Christianity and critique of the Neolithic Revolution in WMS). However, as long as we are able to recognize a gap between the world we seek and the world we create and maintain that it can be bridged, we have a vision for change and a source of hope.

The central question of this work is whether an academic investigation into the dominance, oppression, and exploitation by humans over ourselves and the planet could be of any utility. The task, within a colonized institution, on occupied territory, using the master's tools, to envision a future that may be unimaginable from our current vantage point, with any hope to bridge the gap between that vision and its fruition, sounds impossible. Upon much reflection, however, I have concluded that this is no deterrent at all. There simply is no acceptable alternative but to attempt the project of restoration from every possible angle, including efforts within the structures and epistemologies of domination. As Einstein (1946) concluded:

> Science has brought forth this danger, but the real problem is in the minds and hearts of men [*sic*]. We will not change the hearts of other men [*sic*] by mechanism, but by changing our hearts and speaking bravely.... When we are clear in heart and mind—only then shall we find courage to surmount the fear which haunts the world.

Thus, my aim through this project is not to argue specific ideology but to investigate how we may become clear in our hearts and minds. I aim to make what is intuitive and knowable in every human epistemology undeniable in the epistemology of WMS. This intuitive knowledge relates to the ways we interact with one another and with the planet. It is a need for justice as a matter of survival. We can, will, and must achieve a greater potential if we are to remain a viable species on planet Earth. We face crisis, but we have not yet crossed the threshold into oblivion. We must collectively engage in a process of restoration.

The project of restoration is fundamentally one of hope. I am deeply inspired by accounts of rescuers who subjected themselves and their families to unimaginable risk in order to save the lives of victims of the Holocaust (Oliner & Oliner, 1988). Oliner and Oliner (1988) described the importance of hope for those rescuers:

> If we persist in defining ourselves as doomed, human nature as beyond redemption, and social institutions as beyond reform, then we shall create a future that will inexorably proceed in confirming this view. Rescuers refused to see Jews[4] as guilty or beyond hope and themselves as helpless, despite all the evidence that could be marshalled to the contrary. They made a choice that affirmed the value and meaningfulness of each life in the midst of a diabolical social order that repeatedly denied it. Can we do otherwise? (p. 260)

Although the question is rhetorical, the answer is no. As Paulo and Ana Maria Araújo Freire (1992/1994) wrote, "There is no change without dream, as there is no dream without hope" (p. 91). If we believe that the task of creating a better world is insurmountable, then it will be. But if we strive to be unreasonable people who maintain love, hope, and courage despite all reason and reality, then love, hope, and courage become reality. Such an endeavor is a source of individual fulfillment in its own right, regardless of whether our aspirations are externally realized. The Holocaust rescuers were better able to cope with reality and maintain their humanity than were bystanders, even when their rescue attempts were unsuccessful (Oliner & Oliner, 1988). Furthermore, without the hope to try, there would be no possibility of success at all. Thus, if there is to be any hope at all for a better future, it is the hope that we create ourselves.

The struggle for a better world cannot rest on hope alone. It must be informed by critical understanding of what it is that is happening in the world that we find problematic, inspired by a dream for a better world, and fueled by the knowledge that it is possible. Freire (1992/1994) wrote,

> I do not mean that, because I am hopeful, I attribute to this hope of mine the power to transform reality all by itself, so that I set out for the fray without taking account of concrete, material data, declaring, "my hope is enough!" No, my hope is necessary, but it is not enough. Alone, it does not win. But, without it, my struggle will be weak and wobbly. We need critical hope the way a fish needs unpolluted water. (p. 8)

Philosophic considerations and theoretical critique must be critically informed of dominant notions of reality. We can and must reject reality as enforced and defended by subjective and dominant WMS epistemology, but we can only do so by knowing that reality. A creative and radical struggle for a better world does not become reactionary by its acknowledgment of present conditions. As we dream of a better world we have to react to consequences of this reality. The goal must be idealistic, large-scale structural change, but we must simultaneously be working at an individual level for and with those most severely targeted by oppression. If in our efforts to avoid becoming reactionary we focus entirely on the dream of a better world, we risk ignoring the immediate and unavoidable reality (e.g., food insecurity) of those who are intended to benefit the most from change. Our efforts must constantly center on those who are the most vulnerable, marginalized, and disaffected.

Freire (1992/1994) found hope to be an "existential, concrete imperative" (p. 8) and an ontological need. The concept of hope as an ontological need, as something akin to faith, and as a theoretical or even poetic concept is beautifully inspiring. However, the concrete necessity of hope is cemented in reality for me by knowledge that I owe my entire existence to hope because my paternal grandparents are Holocaust survivors. In an interview conducted by the University of Southern California Shoah Foundation, my paternal grandmother, Rosalie Krzesni, gave her account of life before, during, and after the war. She said that my grandfather and two aunts had been in hiding during the war and survived with the help of Dutch rescuers (Krzesni, 1997). I don't know who these people were or why they did it, but the sentiment expressed by the Oliners (1988) applies regardless. For all I know, the rescuers who aided my own family's survival could be among those interviewed by Oliner and Oliner (1988).

My grandmother was liberated from Bergen-Belsen concentration camp by the British in 1945 (Krzesni, 1997). She stated that she believed that she had survived because she was strong and able to work. She reported the conditions there to have been far worse than those that she had experienced in Auschwitz. She talked about seeing thousands of dead bodies and surviving by eating leaves from the ground. Upon liberation, she immediately fainted, having contracted typhus. My grandparents met after the war; both had lost many family members and their spouses. They gave birth to my father in Holland and then immigrated to the United States in 1954 (Krzesni, 1997).

My grandmother said that she decided to do the interview because it was something that my family had not talked about. She said it had been too painful but she thought that it was important that someone should know what had happened after she was gone. In the end of her interview, my grandmother was asked whether there was anything else that she would like to share about her life and the Holocaust. She said, "Have faith, and that is all that I can tell you" (Krzesni, 1997). Although I never talked with my grandparents about the Holocaust, I was very close to my grandmother and she was one of the most caring, strong, stubborn, independent, and inspiring people I have ever known. I think that the faith and strength that she speaks of are not only religious faith and physical strength, but also faith in ourselves and our ability to persevere. It is my grandmother's hope as well as that of the rescuers that I have carried throughout this project.

I would have likely never discovered my grandmother's interview if my research hadn't led me to the Oliners (1988), which inspired me to investigate the testimonials of Holocaust survivors in addition to the rescuers. There is something profoundly significant about finding myself in my research. I've always regretted that I didn't know much about my family history and wished that I'd taken the opportunity to talk to my grandmother about her experience while she was alive. I think that has been the greatest success that my efforts could have and has moved me more than any academic theory possibly could.

My maternal grandparents were also both heavily involved in the U.S. war effort. They both grew up working on farms in the U.S. South. Having lost the family farm during the Great Depression, my grandfather found work throughout the South as a migrant farm worker before joining the army. At the same time, my grandmother went to work in a shipyard to support U.S. war efforts. Both pairs of grandparents achieved middle-class status in the United States after the war and thus I have lived a life of privilege. I do feel pride in my family history and in the efforts of my grandparents and parents to give their children a better life. But I haven't offered this story to suggest that people who are the targets of oppression just need to pull themselves up by their bootstraps. This idiotic idiom is literally and figuratively nonsense. I know that the opportunities that my parents and grandparents found for upward mobility are not strictly a function of their individual merit. They were upwardly mobile in a context that was conducive to that mobility and those opportunities were not and are not equitable—they were arbitrarily defined by a racial hierarchy.

With the knowledge that I experience such privilege despite the fact that my ethnicity has been targeted by oppression and genocide, I feel a sense of duty to those who continue to be more severely targeted. I am extremely angry at the existence of such injustice. This is a natural and important response to injustice. We should all be angry. However, we must focus that anger and avoid becoming consumed by it. It is in anger that I dive headfirst and headstrong into a struggle for justice and for a better world, but it is hope that keeps me from drowning.

I believe that the best I can do is to acknowledge the problematic context of my work, approach it with humility, and critically hope that in the cycles of action and reflection to follow, I will find a way to have a positive influence on the planet and its inhabitants. This work is nothing more than my own process of reflection and if not coupled with action, it will be a meaningless waste of time that should have been directed elsewhere. However, my hope is that I have developed an understanding that will inform and direct me in the work that I feel that I need to do. If my knowledge is a product of the master's tools, then I hope that I can repurpose that knowledge for liberation rather than domination.

I want to make it absolutely clear that this book is not independent of my personal values. There is no avoiding the fact that it is representative of my own agenda for change. This is not an esoteric project aimed at understanding some obscure phenomena, but rather an effort to mobilize existing knowledge in a struggle for a better world. That world, as the Zapatistas say, is a world where many worlds fit. Thus, I have tried to leave much of the definitions of change and of progress open ended and focus more on how and why people engage in such efforts in order to help more people become engaged in their own way.

I have tried to share the perspectives of scholars (both academic and nonacademic) of different disciplines and also different ethnic, cultural, and racial backgrounds, identities, and experiences. I have tried to balance my use of empirical knowledge produced by WMS and "the master's tools" with knowledge of lived experience offered by nonacademic scholars who, by choice or circumstance, have concentrated their efforts on a close reading of the world rather than a close reading of the word. In this effort, I have tried to quote directly rather than paraphrase the work of others as much as possible. Although this work is a representation of my own voice and self-conception, I have conceptualized it as a conversation between the profound thinkers that I have had the pleasure of knowing directly or through written and recorded words. In this conversation, I certainly comment and moderate, but my primary goal is to be an active listener.[5]

I know that I don't have the time, skill, or capacity to fairly represent all of the voices who have contributed to my understanding or to hear all of the voices who should contribute to it. It is very important to me that I get it right (empirically and morally) because this project is a representation of me—more precisely, it is a representation of the person I hope to be. However, this project is unfinished and will always remain unfinished. I don't believe that we ever get to get it right, we only get to progress. In that sense, I hope I look back on this in ten years knowing that it was an important step, but that I overlooked and neglected more critical questions or explicitly had parts wrong. If I am able to say that, I will know that I have made progress in my efforts. Finally, I want to begin this work with an apology to anyone I misrepresent, tokenize, fail to represent, or to whom I fail to give credit. I hope that anyone who reads this will provide me the opportunity to understand and make amends for any such mistakes.

This final draft is only the first draft.

ACKNOWLEDGMENTS

I acknowledge the Lummi Nation, Nooksack Tribe, and Yakama Nation, whose land I have occupied throughout the course of this project.

I would like to thank Gene Myers, Veronica Velez, and Kristen French for their inspiration, guidance, and support. I could never have written this book without them. I would also like to thank the entire Education and Social Justice community at Western Washington University. The stories, conversations, and experiences I shared with my peers in the university community have moved me more than they know.

I would like to deeply thank Sarah Detzer, Koreena Haynes, Orlando Garcia, Edgar Serrano, and all of the parents, volunteers, and youth involved with the Kulshan Creek Neighborhood Program. Furthermore, I want to thank the wonderful young people who attended Homework Club at Villa Santa Maria during of my time there. Getting to know and love the kids in these programs fueled this project and caused my passion and motivation to build a better world to grow exponentially.

I would also like to thank Shirley Steinberg for making the publication of this book possible and Sophie Appel, Christopher Myers, Stephen Mazur, Phyllis Korper, Beth Morel, and all of the fantastic staff at Peter Lang.

Additionally, I thank Bandit, my dog-companion, who has spent countless hours patiently watching me read and write, forgoing many walks, and vigilantly reminding me that we need to take breaks to play outside.

Finally, I recognize all of those who have struggled throughout history for justice—especially those who are not recognized in history or by academia. For every iconic hero there are thousands who dedicated their lives to a better world in their own way. I have reflected deeply on how many folks have died during the course this project as heroes, victims, and martyrs in this struggle. Although certainly too many to name, I want specifically to acknowledge Galeano. Galeano was a Zapatista educator murdered by paramilitaries in Mexico for doing exactly the work that I have demanded in this book.

Not a reading of the word alone, nor a reading only of the world, but both together, in dialectical solidarity (Freire & Freire, 1992/1994, p. 105).

INTRODUCTION

> It is written in rage and love, without which there is no hope. (Freire & Freire, 1992/1994, p. 10)

My interest in education is founded in the belief of the fundamental role of education in shaping society. Although I am passionate about sharing and developing knowledge and constructing meaning as an end in and of itself, it is education's potential to catalyze social change that compels me to aspire to be an educator. The premise of this work is that social degradation and ecological degradation are a single anthropogenic process. As we degrade the planet, we degrade ourselves and as we degrade ourselves, we degrade the planet. I argue that because the processes are inseparable, any activism or educational effort for the benefit of the natural environment must deeply address social justice, and any activism or education toward social justice must significantly include the natural environment.

What we know is no more significant than the way we know it, and the way we know something is inseparable from ourselves and our environment. To deeply know something tacitly rather than just memorize explicit facts—to understand something and incorporate it into ourselves—we must experience it. We must connect our cognition, affect, and the environment in which knowledge exists. Knowledge has place, time, social, and emotional context and it is conditional. I seek to understand the conditions of knowledge that may cause degradation or support restoration. By understanding the conditions that produce ecological and social degradation, I hope to design pedagogy for restoration.

This work is intended to bridge the wisdom of environmental education (EE) and social justice education (SJE) and apply ecological and social justice concepts to the broader process of education. Neither EE nor SJE can be discrete content areas. They should be themes fully integrated into all facets of education; justice for the natural environment and for people must be a primary concern in all public and private life.

Hopeful Approach

My work is distinctly theoretical; while empirical research reinforces many of the concepts, it is apparent that theory dominates. The potential for confirmation-bias and overgeneralization in research based in theory is high (Greenwald, Pratkanis, Leippe, & Baumgardner, 1986). To address this, I have tried to be clear about the values and assumptions that define this work. The central assumptions of my work are that social and environmental degradation has occurred and that oppression exists. I am less interested in providing evidence to support such realities (that I have accepted as truth) than in understanding the conditions producing degradation and designing strategies to create a different reality (pedagogy for restoration). To design strategies for pedagogy for restoration, I am trying to understand under what conditions degradation prevails. Condition-seeking and design strategies are described in "Under What Conditions Does Theory Obstruct Research Progress?" (Greenwald et al., 1986).

A central assumption in this work is that ecological and social degradation is affected anthropogenically. In the search to understand how and why degradation occurs, I will make no prefatory attempt to catalog the myriad manifestations of the human-caused ecological and social crisis, prove that we are in a period of mass extinction, or that climate change is occurring much more quickly than can be explained by natural processes.[1] "Awareness of consequences and information related to land use and social conflict" has been shown to affect moral reasoning about the environment (Berenguer, 2010) and

> such issues require urgent attention, [but] basic knowledge about them does not lead us to a clear path to action, nor does it motivate participation in their solution. Rather, such all encompassing negativity often leads to feelings of powerlessness, apathy, guilt and disillusionment, clouding the path towards real solutions. (Tilbury, 2007, p. 124)

Advocating and analyzing the specific environmental issues that we face today is not a task to be dismissed or trivialized. However, these issues are symptoms of larger processes that I hope to examine. I strive to embrace both/and[2] rather than either/or logic and thus believe that both the specific environmental issues and systems of production of those issues must be addressed. The specifics have been and are continuing to be thoroughly researched, so I hope to make a contribution through a broader theoretical treatment. For the purposes of this work, acceptance (or openness to the idea) that social and ecological degradation is significantly caused (or even affected) by humans is sufficient.

There seems to be a common structure of argumentation that begins with a shock-and-awe approach to emphasize the importance of an issue and concludes with some shred of hope. This is not unlike the structure of the Preface to this book, but for the remainder of this work, I attempt a hopeful approach. Harré (2011) suggested that there may be an optimal ratio between offering positive solutions and identifying the severity of environmental problems of between 3:1 and 6:1. Obviously this is difficult to quantify and attempts to do so have been heavily criticized (Brown, Sokal, & Friedman, 2013), but the point is that humans are likely to be more motivated to solve today's problems by positive action-oriented solutions than dwelling on the associated threats.

There may be an evolutionary basis for a positive psychology based in our fight-or-flight response (Haidt, 2006). We are evolutionarily disposed to react much more instantaneously and intensely to immediate threats than we are to opportunities. This would be a necessary survival strategy because missing a single opportunity generally would not threaten our life, but failing to respond to a single threat, such as a predator, would likely result in death. Our fight-or-flight response served us well in the past and is important in certain situations. However, in the modern world where threats are often abstract and persistent, we are poorly adapted for coping with sustained stress and often ignore or deny a threat or become overwhelmed by negativity (Haidt, 2006). We frequently see both reactions in response to climate change.

My position, as taken by Harré (2011), is that an understanding of the severity of a problem is important to inspire action, but too much information may lead to hopelessness and inaction. I don't think this is something that WMS can precisely quantify, but we can take it as a matter of folk wisdom that we need to adaptively balance positivity and negativity and that neither extreme is healthy (e.g., denying that any problems exist, or feeling doomed). Thus, I will not shy away from the problems in my critique, but neither will I intentionally sink into the muck of hopelessness. I am striving for a balance between critique and hope that is referred to as critical hope.

> Critical consciousness...involves not only a critical social analysis of inequities and actions taken to achieve greater social justice, but also a sense of personal efficacy or perceived capacity to effect change. Some people, for instance, may be aware of systemic inequalities, but may be unlikely to take action to improve systems unless they believe that their actions will likely be effective. The achievement of critical consciousness therefore requires people to attain an understanding of structural and historical forces that impinge on and constrain collective action for social justice, yet to simultaneously maintain their hope that—despite these forces—their own participation in justice-focused action can be effective. (Christens, Collura, & Tahir, 2013)

The challenges that humans and the rest of the planet face are undeniable but the problems identified in society and in education present opportunities for improvement.

Finally, in this effort to maintain a hopeful approach it is imperative to focus on assets, solutions, and opportunities on the path to restoration rather than on deficits, problems, and barriers in a fight against oppression. Often, the more attention an author dedicates to identifying a problem, the weaker the proposed solution will be. Much of the works I've studied importantly problematize or deconstruct existing social structure, knowledge, or epistemology, but provide little direction toward solutions. A great solution doesn't always need to be prefaced by a long problem statement, theoretical critique, or philosophical deconstruction. This project is about constructive and creative pursuits focused on what could be rather than deconstructive and overly critical accounts of reality.

Structure of the Book

Part I addresses the status quo to set the stage for my investigation into human empathy and morality toward a pedagogy for restoration.

Chapter 1 defines environmental education (EE) by establishing a historical dictate through examining its founding documents. This exploration reveals a broader interpretation of EE, calling for greater attention to social issues and urban environments. Additionally, the exploration reveals that EE was intended to be an approach to education reform emphasizing participatory action in solving broadly defined environmental problems such as poverty. Next, Chapter 1 investigates the current practice of EE and the context in which EE takes place with special attention to urban environments. The chapter concludes by suggesting how EE may move forward in consideration of its historical dictate and present context. Ultimately, the chapter attempts to frame EE as the foundation for pedagogy for restoration.

Chapter 2 investigates how and why oppression happens. The chapter investigates how the ways we discuss oppression in the work of anti-oppression may undermine our efforts. Next, various theories on oppression are examined using the Marxist critique of capitalist production and distribution of goods as a foundation. Epistemological and ecological perspectives on the roots of oppression are also considered. I suggest that although it may be important to consider how and why oppression happens, it may not offer significant utility toward ending oppression. The chapter concludes with a consideration of restoration as an opposing force to oppression.

Part II attempts to understand human morality as a motivator for prosocial and pro-environmental behavior. The emphasis of Part II is on understanding the human potential for *goodness* and how we are motivated to do *good* for one another and for the natural environment.

Chapter 3 explores human empathy. The chapter begins by exploring how empathy and morality may have evolved to be human dispositions as well as potential pitfalls of such an approach. Next, empathy and its development are explored as precursors to morality. Finally, the limitations of empathy are examined, which begins to shed light on why our empathic disposition does not result in perfect harmony among humans.

With a solid foundation in empathy, **Chapter 4** continues by exploring human morality, how it develops, and the specific principles of rights, caring, and justice that we may consider in the process of determining right from wrong.

Chapter 5 examines social influences and individual differences that affect our moral judgments and actions. Moral limitations and conflict are examined, specifically considering how we exclude individuals or groups from moral consideration and differing conceptions of right and wrong. Next, outlying behavior is examined with a sympathetic approach to understanding how and why some may behave in a harmful manner, emphasizing the role of stress and trauma. The chapter concludes by exploring how people are triggered by certain conflicts to engage in prosocial and pro-environmental behavior and outlines a variety of ways in which individuals are motivated.

Part III significantly departs from the previous chapters and attempts to detail elements of pedagogy for restoration. I propose that pedagogy must concentrate on helping students meet their physical and emotional needs and encourage a sense of efficacy. I suggest that learning, growth, and morality will follow if the environment is conducive to allowing them to do so. This approach to education is composed of three elements: pedagogy of basic needs, pedagogy of identity, and pedagogy of purpose.

Chapter 6 proposes a pedagogy of basic needs and suggests that meeting students' basic physiological and psychological needs is prerequisite to all educational efforts. I suggest that meeting students' basic needs is morally imperative but that meeting needs can also be incorporated into a meaningful educational experience. The chapter takes hunger as an example and proposes the ways in which a pedagogy of basic needs could reframe the discourse of need and transform a classroom into a community development organization whose purpose is to eradicate hunger.

Chapter 7 proposes a pedagogy of identity and focuses on facilitating learning relevant to students' own experiences. A pedagogy of identity is equally focused on self-concept and relationships, employing multicultural education intended to honor our similarities and our differences while emphasizing that we are all connected. Hip Hop is explored as an American urban culture relevant to many students' experiences with examples of how Hip Hop facilitates education.

Chapter 8 proposes a pedagogy of purpose and focuses on helping students develop a sense of purpose in the world to engage in their own way and on their own terms. This pedagogy explores how a sense of purpose develops and how educators can support students in that pursuit. The chapter concludes by coming full circle and illuminating that each element of pedagogy supports and reinforces the other.

In **Chapter 9,** I revisit the practice of environmental education and consider how pedagogy for restoration may inform or be informed by environmental education.

The work concludes with a discussion of my own experience writing this book, the limitations of my approach, and—most important—a demand for action.

PART I: THE STATUS QUO

· 1 ·

ENVIRONMENTAL EDUCATION

> Environmental education should promote the strengthening of peace, the further relaxation of international tensions and mutual understanding among States and be a real instrument for international solidarity and for elimination of all forms of racial, political and economic discrimination. (UNESCO, 1978, p. 26)

Historical Dictate of Environmental Education

To establish the foundation of this work in environmental education (EE), I examine the foundation of EE. Although education in and about the environment has almost certainly taken place in some form for as long as humans have existed, the thrust of the EE field was defined in the 1970s. The first major document was the Stockholm Declaration in 1972, a result of the United Nations Conference on the Human Environment, which aimed to address "the need for a common outlook and for common principles to inspire and guide the peoples of the world in the preservation and enhancement of the human environment" (UNEP, 1972). The Stockholm Declaration identified an educational need which the Belgrade Charter attempted to address in 1975 (UNEP, 1975). The Belgrade Charter established the framework and guiding principles for EE, which were further developed at the United Nations

Educational, Scientific, and Cultural Organization (UNESCO) Intergovernmental Conference on Environmental Education, held in Tbilisi in 1977.

The officers of the Tbilisi conference commission represented the USSR, Belgium, Benin, Colombia, Czechoslovakia, Egypt, Finland, Iraq, Kenya, Mexico, Thailand, Canada, India, Bulgaria, Malaysia, Sierra Leone, the United States, Venezuela, and Tunisia (UNESCO, 1978). The Tbilisi Declaration is a condensed summary of the conference proceedings and provides the most commonly used definition of EE. The definition is often paraphrased and a thorough reading of the Final Report on the conference proceedings provides much broader conceptions of EE than such definitions usually reflect.[1] The Tbilisi Declaration (1978) stated:

> Environmental education, properly understood, should constitute a comprehensive, lifelong education, one responsive to changes in a rapidly-changing world. It should prepare the individual for life through an understanding of the major problems of the contemporary world, and the provisions of skills and attributes needed to play a productive role towards improving life and protecting the environment, with due regard given to ethical values. By adopting a holistic approach, rooted in a broad, inter-disciplinary base, it recreates an overall perspective which acknowledges the fact that natural environment and man-made [*sic*] environment are profoundly inter-dependent. It helps reveal the enduring continuity which links the acts of today and the consequences for tomorrow. It demonstrates the inter-dependencies among national communities and the need for solidarity among all mankind [*sic*]. ("The Tbilisi Declaration," 1978)

The focus of environmental education was not strictly defined to be the natural environment. The emphasis was on the sustainability and quality of life as both a socially and ecologically focused effort. According to the Belgrade Charter and Tbilisi Declaration, EE is about the relationship between humans and the environment and is an interdisciplinary approach rather than a narrow ecological one.

Economic development is a central concern in each of the three documents. The Executive Director of the United Nations Environment Programme opened the conference by calling for the "sustained improvement of the quality of life, eradication of poverty and equitable participation of people in the benefits of development" (UNESCO, 1978, pp. 61–62).

> Environmental problems are not just those of the detrimental or irrational use of natural resources and pollution. They include problems of underdevelopment, such as inadequate housing and shelter, bad sanitary conditions, malnutrition, defective management and production practices and, more generally, all problems which stem

> from poverty. They also include questions of protecting cultural and historical heritages. (UNESCO, 1978, p. 11)

Thus, "poverty itself is a form of environmental degradation" (UNESCO, 1978, p. 11).

Although economic development was considered imperative, "unprecedented economic and technological progress" was recognized to bring "inevitable changes, improvements, but also environmental hazards" (UNESCO, 1978, p. 40). In fact, this conflict between human interest, economic interest, and the well-being of the planet was the central concern of the Stockholm Declaration. The first paragraph of the declaration stated:

> Our generation has witnessed unprecedented economic growth and technological progress which, while bringing benefits to many people, have also caused severe social and environmental consequences. Inequality between the poor and the rich among nations and within nations is growing; and there is evidence of increasing deterioration of the physical environment in some forms on a world-wide scale. This condition, although primarily caused by a relatively small number of nations, affects all of humanity. (UNEP, 1972)

The Belgrade Charter (1975) was prefaced with a statement in direct response to the Stockholm Declaration calling for a new international economic order,

> one which takes into account the satisfaction of the needs and wants of every citizen of the earth, of the pluralism of societies and of the balance and harmony between humanity and the environment. What is being called for is the eradication of the basic causes of poverty, hunger, illiteracy, pollution, exploitation, and domination. The previous pattern of dealing with these crucial problems on a fragmentary basis is no longer workable. (UNEP, 1975)

In each of the three documents the emphasis is on development that protects human rights and improves human welfare in a manner that is not destructive to the natural environment. Such a concern for detrimental influence of economic interests on both human welfare and the natural environment and a vision for positive development was the premise of environmental education.

The goal of EE was to "ensure the preservation and improvement of humanity's potentials and develop social and individual well-being in harmony with the biophysical and man-made [*sic*] environment" (UNEP, 1975). This conception of EE certainly did not arbitrarily identify ecological degradation as an environmental issue and social concerns as separate humanitarian ones. In fact, it was meant to bridge the two areas of concern and position the needs of humans and the needs of the planet in congruence through a new global ethic and educational reform (UNEP, 1975): "To improve all ecological relationships, including the relationship of humanity with nature and people with each other" (UNEP, 1975). This new ethic was intended to be global, but not static or universal. The relationships were meant to be interpreted on an individual basis in accordance with cultural context.

Environmental education was not about specific learning objectives and was neither intended to be peripheral to public education nor an addendum to it. Environmental education was meant to reform all public education. The Belgrade Charter (1975) explicitly stated the necessity of educational reform:

> The reform of educational processes and systems is central to the building of this new development ethic and world economic order. Governments and policy-makers can order changes, and new development approaches can begin to improve the world's condition—but all of these are no more than short-term solutions, unless the youth of the world receives a new kind of education. This will require new and productive relationships between students and teachers, between schools and communities, and between the education system and society at large. (UNEP, 1975)

The Tbilisi Declaration was largely consumed with expanding on the broad and sweeping demands of the Belgrade Charter and began to elaborate on how exactly EE might reform education.

> Environmental education must look outward to the community. It should involve the individual in an active, problem-solving process within the context of specific realities and it should encourage initiative, a sense of responsibility and commitment to build a better tomorrow. By its very nature, environmental education can make a powerful contribution to the renovation of the educational process. ("The Tbilisi Declaration," 1978)

The Tbilisi Declaration clearly defined EE as a community project aimed at improving the lives of human and all living beings and a critical approach to reforming the educational process.

All three documents were much more focused on the approach to EE than the specific practices, intuitively aware that specific application could not be context independent. Although certainly implicit, specific ecological knowledge was not emphasized in the Tbilisi Declaration (1978) or the preceding documents. Rather, the Tbilisi Declaration (1978) emphasized general understanding and knowledge necessary for participation in problem solving. The UNESCO Report envisioned an inclusive environment where

> Active participation would...replace intellectual conjecture. Moreover, the varied origin of participants, which, in conventional education might be regarded as a difficulty, would then become an advantage, since it would be possible to achieve the spontaneous integration of different disciplines and different ways of thinking in a joint experience directed towards the solution of problems. (UNESCO, 1978, p. 22)

Without explicitly using such terminology, Tbilisi and Belgrade screamed for popular educational approaches demanding reform such as integrative education, participatory action research, and critical pedagogy. In summary, "environmental education must adopt a holistic perspective which examines the ecological, social, cultural and other aspects of particular problems" (UNESCO, 1978, p. 12).

Practice of Environmental Education

The early vision for EE detailed in the UNESCO Final Report and Tbilisi Declaration is quite radical in consideration of the practice of EE and environmentalism today. Gruenewald (2004) calls EE "politically sanitized" and suggests that EE should "not be content with the self-inflicted narrowness that the adjective [environmental] inevitably conveys" (p. 101). The North American Association for Environmental Education (NAAEE) "adopts language and concepts associated with conventional standards: grade-level performance, achievement, and effectiveness" (Gruenewald, 2004, p. 80). Alignment with conventional standards may have improved the credibility and receptivity of EE and increased its reach and presence in mainstream education. However, by such alignment, EE may have been reformed and disciplined by conventional education rather than reforming education as it was intended to. Gruenewald (2004) suggested that alignment with standards and institutionalization of EE "serve to legitimize, rather than challenge, educational practices that are problematic, that in fact create the need for environmental education" (p. 76). Such alignment of "EE with the official, high-status content-area standards of general education has become a trendy means of legitimization, one that may work against transforming the ecologically problematic thrust of general education" (Gruenewald, 2004, p. 80). The alignment of EE with economic values of "individualistic competition and nationalistic success in the global economy... rather than emphasizing that EE calls for a new (or ancient) ecological way of knowing" (Gruenewald, 2004, p. 80) may, indeed, be a problematic thrust significantly contributing to social and ecological degradation.

The narrowness and political neutrality that Gruenewald (2004) criticizes in EE (whether or not it is accurate) are not dictated by the Belgrade Charter or Tbilisi Declaration. The UNESCO Report calls for EE to "consider the environment in its totality—natural and built, technological and social (economic, political, technological, cultural-historical, moral, aesthetic)" (UNESCO, 1978, p. 27). I broadly conceptualize the environment as the space in which life exists. This space is equally subject to and constructed by ecological and social forces from climate change to gun violence. Henceforth, I intend this broad interpretation of the word "environment" unless otherwise specified. Furthermore, all education takes place in the environment, is informed by the environment, and applies to the environment, so all education is environmental education. Thus, my use of the word "education" is intended to reflect the broad process of education that takes place inside and outside of traditional classrooms. The broad process of education certainly includes the specific practice of EE today and concept of EE described in the UNESCO Report.

Context of Environmental Education

The case has been made for a broad concept of EE, based on its founding documents and as a mostly theoretical pursuit. Now I examine the social environment in which EE takes place. Garbarino (1996) described the environment in which many people in economic poverty live as both socially and physically toxic.

> Poor children live in the kinds of socially toxic environments that generate multiple threats to development—academic failure, child maltreatment, learning disabilities, and others. That is one clear meaning of being poor in America. Interestingly, this social toxicity parallels physical toxicity; low-income populations are more likely to be exposed to chemical and radioactive waste and polluted air and water. (Garbarino, 1996)

EE must attempt to make restoration of the social as well as the ecological environment more salient. In a study on environmental perceptions in a region with high levels of economic poverty, Kahn (1999) interviewed children about their perceptions of the environment in a Houston, Texas, school. The school was attended predominately by Black[2] students (99%); most students received free and reduced lunches (91%) and the majority of the children were labeled as low achieving (60% based on state tests). One student was asked a preliminary question about a bayou that flowed within a mile of the school:

> Tell me Trina, do you know what a bayou is? *Yes…it's where turtles live and the water is green because it is polluted. People—some people need to, um, some people are nasty. Some people, you know, like some people go down there and pee in the water.* Mm hmm. *Like boys, they don't have nowhere to pee, and drunkers, they'll go do that, too.* Okay. *And sometimes they'll take people down and rape them, and when they finished, they might throw 'em in the water or something.* So, what does it look like? How would you describe it? A bayou? *It's big and long and green and it stinks.… And turtles live in it.* (p. 96)

Trina was in the third grade. It is clear that her basic knowledge of the ecosystem was entirely interwoven with the realities of the city. I revisit this study as it relates to moral judgments in other chapters, but here, even if merely anecdotally, I wanted to illustrate that the distinction between social and ecological problems is arbitrary. It is from a position of significant privilege that one could suggest that, as an environmentalist, certain social issues are not within the scope of their concern. I doubt Trina had the luxury to dismiss concerns about rape and murder to focus attention on turtles, for which she did seem to have some affinity.

An interesting result from a complementary study in which Kahn (1999) interviewed the Houston children's parents was that 57% of parents considered drug education and environmental education to be equally important and 29% of parents considered environmental education to be more important. I wonder whether the curricula of drug education programs might have influenced the parents' judgments. However, one parent explained that drug problems and environmental problems both threatened lives in their community. For that reason, they found environmental and drug education equally important.

> *With the drugs, we're nothing. Without the environment, we're nothing. And drugs is something I see every day. There are dealers across the street from me. So I see this every day and it's just killing us. I mean, it really is killing us, and with the drugs we're not going to have any youth.… With the drugs you're not going to have a future and without any environment you're not going to have a future.*
>
> *Well let's put it like this here. If you don't take care of one* [drugs], *it's going to kill you. If you don't take care of the other* [the environment] *it's going to kill you.* (Kahn, 1999, p. 122)

To further contextualize, parents described garbage piling up in the streets, sewage backing up into their yards, and the prevalent stench of air pollution.

In this example, I have tried to illustrate that in the daily lives of many (most) folks there is little distinction between the social and natural environment. I have also tried to make a case for a broader conception of EE (with greater attention to social issues) by dictate (UNESCO Report). And in the following chapter, I make a theoretical argument for the relationship between social and environmental degradation.

Moving Environmental Education Into the Future

> In the modern era, economic growth and development have always depended on exploiting natural and human resources for the sake of economic production and the creation of wealth. Such is the history of nation building: imperialism, colonization, cultural violence, and ecological rape.[3] Growth, development, resources, production, wealth—all of these metaphors have taken on positive value in modern culture, yet each is linked to destructive and oppressive relationships with human and nonhuman others. (Gruenewald, 2004)

It is crucial to understand where knowledge accepted as truth comes from and how it has changed over time. Knowledge that we take for granted may, despite our intentions, carry with it a history of degradation that we subtly drag forward by not interrogating the roots and assumptions of that knowledge. However, Tbilisi provides a definition of EE satisfactory for the project of restoration. I was certainly pleasantly surprised by the content of EE's founding documents, with which I assumed a certain level of familiarity (without having read them) based on how prominently they are referenced by others. As I discovered, the foundations of EE—upon which I construct a pedagogy for restoration—are very supportive of such an effort. But while it is important to examine the past and present definitions of EE, there is no reason to remain constrained by those definitions. As Noam Chomsky frequently reminds us, the world is too complex for history to ever reproduce itself. Context changes and history is not static, so our approach to restoration must always be adaptive. I very decisively look to the imaginary to chart my own path into the future with an understanding that the task would be impossible without knowing the past and present context of education and of the issues that must be addressed.

I critically and hopefully look to the future but not to change for the sake of change or to blind progress. Sandy Grande, an Indigenous scholar who identifies as a Quechua woman,[4] sharply critiqued whitestream[5] education's influence on American Indian sovereignty and identity and discussed the fallacy of "belief in progress as change and change as progress" (2004, p. 69). Grande provided nuance to the role of Indigenous knowledge in education by reminding us that Indigenous knowledge is not historical and static, but that change isn't universally progress. She suggested that there is much to learn from the past and the present but never denied the need to look to the future. What Grande (2004) insisted was that we look critically and holistically to past, present, and future in an attempt to transcend the confines of modernism, capitalism, and colonialism without tokenizing or dismissing traditional and contemporary Indigenous knowledge. We must carefully consider the relationship among place, culture, and epistemology with the knowledge that Indigenous culture and epistemology has forever been inseparable from place, as has been the case for most ways of knowing and being throughout human history but is not the case for dominant Western culture.

There is a further interpretation of the fallacy that Grande (2004) identified that needs to be considered. The juxtaposition of the impact that the original inhabitants of the land currently occupied by the United States have had since time immemorial with the immense social and ecological degradation that the "progress" introduced by white settler colonialism of the past few hundred years paints a certain picture. It seems that the obvious question that such a juxtaposition raises is, What would the United States look like if it had never been colonized by Europeans or if the relationship between European settlers and Indigenous peoples had developed differently? It seems trivial to speculate either that it would be a perfect utopia in harmony with nature or that a similar trajectory of technology, domination, and destruction would have emerged and equally degraded the land.[6] In my mind, it is not trivial, however, to examine counterexamples to the modern and imagine alternative trajectories.

In constructing a pedagogy for restoration, the relation between past, present, future, and imagination must all be continually addressed in relation to one another. Thus, I have begun this chapter by looking to past concepts of EE, present concepts of EE, and how that plays out in the context of U.S. poverty in order to understand how we can begin looking to the future. Such poverty is a symptom of social degradation that must be addressed through pedagogy for restoration. In the remaining chapter of Part I, I attempt to understand the nature of degradation as both an outcome of dominant Western epistemology and social structure. In Part II, I examine how we develop our sense of morality in the context of degradation in order to understand how and why we take restorative action. Finally, in Part III, I explore a pedagogy for restoration.

· 2 ·

OPPRESSION

Even before I ever read Marx I had made his words my own. (Freire, 1998, p. 115)

In this chapter I attempt to understand, at a structural level, how and why oppression happens. I begin by considering discourses of oppression and the implications of the oppressor-oppressed dichotomy. Next, Marxism is briefly introduced as a central theory on the nature of human oppression and as a foundation upon which to investigate other social theories. Following Marx, I delve into Horkheimer and Adorno's (1944/2002) critique of the modes of production of knowledge that raises fundamental questions of the relationship between epistemology and oppression. Then ways of knowing in Indigenous and Eastern philosophies are considered in search of a way to proceed beyond the deconstructionism that critical theory inspires. Finally Bookchin's (1982) theory of human ecology positing that oppression is the result of a departure from nature is investigated. I conclude the chapter by trying to put together social theory, epistemology, and social ecology in order to move beyond an effort to understand the roots of oppression toward an understanding restoration.

Discourse of Oppression

Before proceeding with this investigation into oppression, I want to consider the importance of language and discourse. Often, the way we say something is no less significant than what we say. Although our words don't always match our intentions, if we're not precise in our use of language, we are not precise in our thinking. In this section, I consider discourses related to oppression and how language aligned with the status quo may detract from efforts to challenge oppression.

Oppressor-oppressed dichotomy

By calling people oppressed, without carefully considering what that implies we may subtly validate oppression and undermine our intentions to fight oppression. The oppressor-oppressed dichotomy labels some as oppressors and some as oppressed (or colonizer-colonized, dominator-dominated). The oppressor-oppressed dichotomy isn't an accurate descriptor of status or identity for most people. There are people who enact oppression and people who are targeted by oppression. They are first and foremost people and their identities do not reduce to oppressor or oppressed. Most people experience both to some degree but the significance of either experience to one's experience or identity certainly varies.

If people targeted by oppression are referred to as if it is part of their identity, it subtly justifies their mistreatment by implying that their mistreatment is dictated by their identity. An example may make the point more clear. If someone is referred to as a bum, hobo, or even with a more life-affirming term such as homeless person, we are not only communicating that the person does not have a home (as defined by the status quo), we are also communicating that they are the *type of person* who is homeless. If we refer to someone without such labels, but as a person who does not have consistent access to adequate shelter (or may choose not to use it for various reasons), we make it clear that homelessness is not necessarily part of the person's identity, but part of that person's experience. By avoiding such essentializing labels and seeing a person having a particular experience, rather than someone whose identity dictates that experience, we afford them the basic human dignity of self-determination. Having inadequate shelter is an experience that may or may not be transient and one's own (or perceived) agency in the matter could be determined only on an individual basis. We should not assume that "homeless" people need or want rescuing. Neither should we assume that they will inevitably be "homeless" because it is part of who they are.

Inflicted labels of identity are easily internalized and refusing essentialist concepts of identity allows individuals experiencing homelessness to speak back to their experience rather than to be defined by it. Thus, support efforts can be undermined from the start by the way we define the identity of those we aim to support. I see the task as one of ensuring that all people have access to basic shelter, support, and services that they may need while allowing them the freedom to choose how, when, and whether they use those resources. When we aim to help people while inflicting and enforcing labels of identity, defining their experience and their needs, and presuming that they need or want our help, we may unintentionally maintain an imperialist discourse of oppression that negates our well-intentioned efforts by maintaining the status quo and disempowering "oppressed peoples."

Entitlement and deservingness

Lisa Marie Cacho (2012) similarly discussed how discourse on justice may undermine work for justice. Cacho's areas of concern are primarily the relationship between identity, status and rights. She suggested that those identified as illegal immigrants, gang members, or terrorists are considered ineligible for rights based on assumptions of their status, regardless of their actions. Cacho suggested that by arguing that people deserve rights by distancing them from such statuses, we comply with a racist ideology that suggests some people should have to prove their deservingness for rights while others are entitled to rights. For example, "These people are not criminals; they are just trying to earn a living" is a frequently expressed sentiment intended to illustrate categorical deservingness of rights for People of Color. However, expressions such as "Not all white people are criminals" are so universally accepted that they don't need to be said because whites[1] are entitled to rights. This is not to say that there are not identities and statuses among whites who are also unjustly forced to prove their deservingness of rights, but that the rights of whites are not categorically called into question. The baseline should be that everyone deserves rights and that rights should not be contingent on status or identity.

> When we distinguish ourselves from unlawful and outlawed status categories, we implicitly insist that these socio-legal categories are not only necessary but should be reserved and preserved for the "genuinely" lazy (welfare recipients), "undoubtedly" immoral (marrying for citizenship), and "truly" dangerous (gang violence). When we

> reject these criminalized others of color, we leave less room for questioning why such status categories are automatically and categorically devalued. While these tactics may be politically strategic and even necessary at times, it is important to be cognizant of the fact that they work because a sympathetic public can register that some people are the wrong targets of legitimate laws. They work only if a sympathetic public already accepts that discrimination against not-valued others is legitimate and necessary. (Cacho, 2012, p. 18)

These sorts of comparisons are symptomatic of an unjust dichotomy where some folks are entitled to rights while others are required to prove their deservingness of rights. By making implicitly comparative arguments of race and relation and insisting that one person or group deserves rights we are implying that some other doesn't.

However, as Cacho (2012) pointed out, it is sometimes practically necessary, and even unavoidable, to make such relational arguments. For example, in immigration law, and in the United States more generally, heteronormative conceptions of family are highly valued (Cacho, 2012). For example, immigrants facing deportation are often portrayed and valued in accordance to heteronormative family values (e.g., the father struggling to earn money to send home to his family, or the mother facing deportation, which would tear apart her family). Certainly, I am not in support of tearing apart families and I think it is a noble effort for someone struggling to make ends meet to send money home. Thus, advocacy and activism related to immigration reform often (maybe unavoidably) emphasize family values because they are likely to be effective. However, such arguments are likely to be effective at the expense of less-valued others (e.g., single, transgender refugees). Cacho (2012) ultimately did not argue that, in practice, such arguments are ineffective or entirely problematic, but that we must be extremely cognizant of the ways the language of social justice may undermine social justice. "What's at stake is figuring out the criminalized and racialized parameters of rights discourses by realizing that the ways in which a group's demands for rights and recognition can highlight an/other racialized group's ineligibility for those same rights" (Cacho, 2012, p. 82). Therefore, what we may ultimately be constructing is a politics of misrecognition "that relies on tactical arguments that construct only some members of a group targeted for state violence as having been falsely and unfairly misrecognized by US law" (Cacho, 2012, p. 28) and more generally targeted and misrecognized in society.

Production of Oppression

I begin my exploration of the mode of production and reproduction of environmental destruction with a nod to the influence of Karl Marx. Marx is one of the influential thinkers of my own worldview and the influence of Marx in the schools of thought of the majority of the authors that I have referenced is easily recognizable (if not overtly stated). However, I have made an effort to avoid an intransigent take on Marxism or any particular ideological allegiance in order to look forward both critically and imaginatively.

Marx's basic presumption was that the class struggle, with its associated oppression, propels history and explains the nature of all social structure. Marx suggested that revolution was necessary to redistribute the means of production that would, in turn, dissolve class structure (Marx & Engels, 1847/2010). Marx certainly expressed great concern for the exploitation of workers and (if only as a digression) expressed some concern for the natural environment:

> The bourgeoisie, during its rule of scarce one hundred years, has created more massive and more colossal productive forces than have all preceding generations together. Subjection of Nature's forces to man [*sic*], machinery, application of chemistry to industry and agriculture, steam-navigation, railways, electric telegraphs, clearing of whole continents for cultivation, canalisation of rivers, whole populations conjured out of the ground—what earlier century had even a presentiment that such productive forces slumbered in the lap of social labour? (Marx & Engels, 1847/2010, p. 17)

Marx identified the owners of material means of production as perpetrators of the exploitation of human beings, but rarely (or never) examined the relationship between the processes of production and destruction of nature. As a vapid critic of the "idiocy of rural life," and a vehement proponent of industrialization, Marx did not consider that the very materialist nature of production may inevitably be exploitative to humans and the natural environment (Marx & Engels, 1847/2010). Thus, Marx may have had some awareness that exploitation of the planet was a problem, but didn't question the nature of a worldview that holds Earth as a resource. Still, Marxist theory does provide the inspiration for the argument that social and environmental degradation is deeply related to social structure and is readily analyzed as a single issue.

Epistemology of Oppression

Following Marxist thought, critical theorists Horkheimer and Adorno (1944/2002) sought to understand the nature of oppression in a time when modern science very prominently enhanced the scale of atrocity in World War II. In the midst of domination of Nazi fascism, Horkheimer and Adorno (1944/2002) concluded that the production and distribution of knowledge rather than capital (as Marx thought) was the fundamental source of oppression.

> Knowledge, which is power, knows no limits, either in its enslavement of creation or in its deference to worldly masters. Just as it serves all the purposes of the bourgeois economy both in factories and on the battlefield, it is at the disposal of entrepreneurs regardless of their origins. (Horkheimer & Adorno, 1944/2002, p. 2)

While Marx critiqued capitalism, Horkheimer and Adorno critiqued modernism. They believed that enlightenment, even in pursuit of liberation, disenchanted the world and that by "conquering superstition" humans established rule over nature. They suggested that "what human beings seek to learn from nature is how to use it to dominate wholly both it and human beings" (Horkheimer & Adorno, 1944/2002, p. 2). Enlightenment established hierarchies of knowledge wherein experts or "masters" held the power. Marx primarily critiqued material distribution in accordance to social relations through capitalism, but not the means of production. Horkheimer and Adorno critiqued both the production and distribution of knowledge. Ultimately, Horkheimer and Adorno felt that knowledge destroyed meaning and enlightenment itself was domination.

Since Horkheimer and Adorno (1944/2002) saw the very concept of progress as domination, they looked to the past and particularly held art as a liberatory form of expression. Horkheimer and Adorno wrote, "The urge to rescue the past as something living, instead of using it as the material of progress, has been satisfied only in art, in which even history as a representation of past life, is included" (p. 25). Rather than suggest that historical knowledge is a source of progress, Horkheimer and Adorno are suggesting that through art, the past lives in the present. In that sense, I think Horkheimer and Adorno saw the study of history (as encapsulated in art) not as knowledge, but transcendence. It may not be too much of a stretch, then, to suggest that Horkheimer and Adorno, amidst a world torn by World War II, found their only escape and fulfillment in the transcendence that one can experience while engrossed in great art. In that sense, their utopia was not externally realizable, but potentially experienced internally, if only briefly.

Horkheimer and Adorno (1944/2002) may be best known for their paradoxical assertion that "myth is already enlightenment, and enlightenment reverts to mythology" (p. xviii). In the context of dominant U.S. cultural mythology, the myth of human nature, the myth of pristine nature, and the myth of socially constructed nature have profound implications that should be considered.[2] In my own experience, and popular among certain environmental educators and writers such as David Sobel, the sort of transcendence that Horkheimer and Adorno may have experienced through art is also satisfied in the wonder of nature (Sobel, 2008). However, it may be that art encompasses our interpretation of nature just as it does our interpretation of history. There may, indeed, be a deep connection between ethics and aesthetics wherein justice is an expression of beauty and beauty itself is justice. In fact, Moore (2005), in an ode to Rachel Carson, concluded:

> There is moral significance in the search for meaning, and virtue in the life of one who seeks—like Carson, attentive and grateful, careful with science and open to mystery, humble and respectful, rejoicing in the fact of things, willing to be surprised. I don't know if humans are the only beings who wonder. But I do know that we have a great capacity to wonder at the world that ticks and sighs around us, and it may be that we will find the fulfillment of our potential as human beings in our awareness of the astonishing world, our care, and our thanksgiving. (p. 277)

I can't help but speculate that Horkheimer and Adorno would dig such a sentiment. My take is that they were so consumed by the atrocity of World War II that the potential to appreciate any element of the present was overshadowed.

I favor a take on enlightenment founded in ancient Chinese and Buddhist traditions. Enlightenment is accepting each moment exactly as it is (Garbarino, 2011). It is "intrinsically and inevitably incomplete and approximate" (Garbarino, 2011, p. 73) because it is an aspiration rather than an outcome. While enlightenment is about accepting each moment exactly as it is, understanding is inevitably subjective. The understanding associated with such enlightenment is more about acceptance, peace, and harmony than it is about domination. The state of enlightenment that Garbarino (2011) described as transformational grace is a sustained state of transcendence quite similar to the transient state that Horkheimer and Adorno (1944/2002) may have found only in art. Garbarino (2011) suggested that transformational grace is a way of being in and with the world where life itself is the source of wonder and transcendence. I like the juxtaposition of Horkheimer and Adorno's concept of (modernist/scientific) objectivity as domination and Garbarino's Eastern concept of objectivity as acceptance. Both views converge in rejection of positivism with acceptance of myth and appreciation of aesthetic.

Garbarino's (2011) thoughts on enlightenment help us out of the hopelessness that Horkheimer and Adorno (1944/2002) may arouse. However, they were correct to criticize dominant Western epistemology and Garbarino (2011) did not speak to structural change. Grande (2004) is certainly very well versed if not significantly influenced by critical theory and postmodern thought, but she's firmly rooted in Indigenous epistemology and ontology, which seems to fuel her vehement cultural critique without the commonly resulting descent into a deconstructionist, fatalist, or nihilist condemnation of every element of human experience and knowledge as meaningless social construct. Grande (2004) identified colonization as the source of oppression rather than capitalism and critiqued modern(ist) colonial modes of knowledge production without assuming that alternative modes of knowledge production do not and cannot exist. She suggested that "both Marxists and capitalists view land and natural resources as commodities to be exploited, in the first instance, by capitalists for personal gain, and in the second by Marxists for the good of all" (p. 27). Her critique of Marxism is aimed not at Marxism in its entirety, but at what she considers a "profoundly anthropocentric notion that presumes superiority of human beings over the rest of nature" rooted in humanism (p. 27). She concludes that "the real source of environmental destruction" is "colonization and the ill-effects of its consuming habits" (Grande, 2004, p. 65).

It seems many of the postmodernists and critical theorists including Horkheimer and Adorno (1944/2002) were stumped upon realizing that oppression is a consequence of the dominant epistemology with little conception of alternatives. Although I found Horkheimer and Adorno's conception of art to be deep and poetic, they didn't have much to offer in the way of positive solutions. As Grande (2004) suggested, it may be that solutions to the problems produced by epistemologies of domination cannot be solved in the same manner of thinking. Horkheimer and Adorno may have lacked the necessary cross-cultural perspective to imagine solutions to the problems that their critique identified. Luckily, as both indigenista[3] and academic, Grande is not bound by the confines of modern Western social theory and identified Indigenous cultures as living counterexamples to the status quo. In this way,

Grande refuses the impossibility of what Horkheimer and Adorno considered theoretically utopian but practically unrealizable in the present: an alternative trajectory.

Ecology of Oppression

Epistemology is culturally influenced and although our way of knowing may be conducive to oppression, it does not explain how the gears were set in motion. What led to the colonial mentality? Bookchin's (1982) work on social ecology begins to chip away at the source of oppression. Bookchin's work followed directly from Marxist thought, but Bookchin's departure from (or advance of) Marxism exposed a fundamental weakness in Marxist theory:

> What puzzled a few highly sophisticated Marxists in later years was how the repression and disciplining of external nature could be achieved without repressing and disciplining internal nature: how could "natural" nature be kept in tow without subjugating "human nature"? (p. 10)

Similarly to Horkheimer and Adorno (1944/2002), Bookchin (1982) saw a relationship between domination of nature and domination of humans, but did not suggest that knowledge necessarily detaches us from nature or causes domination. Bookchin (1982) asserted that "the very notion of the domination of nature by man[4] stems from the real domination of human by human" (p. 1). Bookchin believed that the process begins with patriarchal morality and wrote that

> even before man embarks on his conquest of man—of class by class—patriarchal morality obliges him to affirm his conquest of woman. The subjugation of her nature and its absorption into the nexus of patriarchal morality forms the archetypal act of domination that ultimately gives rise to man's imagery of subjugated nature. It is perhaps not accidental that nature and earth retain the female gender into our own time. (p. 121)

Bookchin (1982) hypothesized that in the process of developing human society, hierarchies among humans were established and that differential class advancement among humans serves as the means to exploit nature. As indicated in the quote above, Bookchin believed that the fundamental source of all oppression is rooted in the oppression of women. However, the development of society represents a departure from a natural order that may precede the oppression of women. It becomes a bit of a chicken and the egg debate. There may be a single root cause of all oppression or there may be many roots.

Finally, as Johnston (1994) concluded in the edited volume *Who Pays the Price? The Sociocultural Context of Environmental Crisis*, surveying international conflicts between ecological health and human rights:

> Environmental quality and social justice issues are inextricably linked. Efforts to protect a "healthy environment" may, in some cases, result in human rights abuse, and depending upon subsequent social response, may ultimately fail to meet original environmental integrity objectives. And conversely, responding to human rights needs while ignoring the environmental context infers temporary intervention rather than substantive solution; it may thus serve to initiate or perpetuate a cycle of human rights abuses. (p. 235)

Johnston's (1994) conclusion is pragmatic and actionable. While it is fascinating to examine whether human exploitation of nature is at the root of human exploitation of humans or vice versa, it may be unanswerable. It may also not be the most productive area of inquiry if the goal is to end oppression. Even if we could be sure that the subjugation of women was at the root of all oppression, that wouldn't necessarily provide much insight toward ending all oppression. However, some understanding of the interconnectedness of the oppression of nature and of humans does provide insight toward the ultimate goal of restoration.

Restoration

> One is not fully human until one acknowledges and affirms the humanity of others—including one's enemies. Ultimately, the enemy is within the human family and not without. And once we acknowledge that, we will all have the courage…[to] move beyond the darkness of mutually destructive hatred and revenge into the light of reconciliation and forgiveness. (Mathabane, 2002, p. A21, as cited in Gibbs, 2014, p. 120)

As Johnston (1994) suggested, our collective struggle against social and ecological degradation must constantly and simultaneously consider environmental quality and social justice issues. At its currently advanced stage, the social crisis and ecological crisis are entirely intertwined, as are development and degradation. Social degradation directly causes ecological degradation and ecological degradation causes social degradation. I hypothesize that oppression, like a virus, infects our way of perceiving the world (e.g., colonial epistemology and internalized oppression) and thus spreads through all of our social structures regardless of where it begins. In the effort for social and ecological restoration, it is important to realize how interconnected different expressions of oppression are. I don't believe that we can make much progress by approaching symptoms of this virus as singular and isolated issues. Ending any oppression requires ending all oppression. Thus we must simultaneously focus on individual instances and structural causes of both social and ecological degradation.

Bookchin (1982) found the end of human domination to be an ecological imperative—"because [human domination] threatens the integrity of organic nature, it will not continue to do so given the harsh verdict of…nature" (p. 37). To be clear, Bookchin was suggesting that feedback mechanisms of the ecosystem will remove or drastically reduce human populations if we do not make significant changes. In my attempt to maintain a hopeful approach, I reframe Bookchin's prophecy. The iniquity that was described in the poem from Egypt four thousand years ago (see the preface to this book) and to which Bookchin similarly referred is not a simple question of moral or amoral behavior but one of great complexity. A large portion of the remainder of this book is devoted to the philosophical, psychological, and sociological exploration of morality. I believe that human concepts of justice have been selected for by evolution and as part of human ecology morality is not detached from the greater ecosystem. If humans establish just ways of living and interacting with one another and with the planet, as I will argue is part of human nature, we will not face the "harsh verdict of nature."

As Horkheimer and Adorno (1944/2002) and later Grande (2004) suggested, the social systems of knowledge production are colonized by the virus of oppression. However, I don't believe that the possibility of transcendent enlightenment is lost. I don't believe knowledge has to be domination but we do need a shift in our concept of knowledge. I believe that there is an epistemology of domination and there is an epistemology of aesthetic, of love, and of association. I want to know the world in the way I know loved ones, where the more I know them, the more affinity I feel for and with them and the more I want to know them. The more I begin to understand their internal states, their needs, their motivations, the more deeply I care. This way of loving knowledge and knowing love is an episteme of beauty and is the vocation of humans. It is a way of being with the planet rather than on the planet. Freire (1998) wrote,

> Women and men [*sic*] by the mere fact of being *in* the world are also necessarily being *with* world. Our being is a *being with*. So, to be in the world without making history, without being made by it, without creating culture, without a sensibility towards one's own presence in the world, without a dream, without song, music, or painting, without caring for the earth or the water, without using one's hands, without sculpting or philosophizing, without any opinion about the world, without doing science or theology, without awe in the face of mystery, without learning, instruction, teaching, without ideas on education, without being political, is a total impossibility. (p. 58)

As Freire said, we are unfinished humans who may become aware of our own unfinishedness and to be human is to search for completeness (Freire, 1998). Such an episteme is generative, but never destructive. It cannot destroy an epistemology of domination but it allows a process of healing.

Our struggles against oppression must begin with processes of decolonizing our own minds (cf. Yellow Bird & Wilson, 2012) because domination is both epistemic and fundamental to our discourse. I remain skeptical of all notions of modern development (e.g., economic development, resource exploitation, technology) as inherently and inevitably a form of domination. However, this skepticism is based in my own tendency toward deconstructionism and is likely symptomatic of a lack of perspective resulting from my own entrenchment in modern Western epistemology. Such skepticism is the hurdle that we must each individually overcome by decolonizing ourselves as part of the effort to end oppression in all of its forms. In this way, pedagogy for restoration is both a process of self-restoration and a broader process of restoration of the planet.

PART II: EMPATHY AND MORALITY

> If we are to live in a world free from the threat of Holocausts, we will need to create it. If we can understand some of the attributes that distinguished rescuers from others, perhaps we can deliberately cultivate them. (Oliner & Oliner, 1988, p. xviii)

Part II explores the complex relationship between the development of an individual's moral principles, the natural environment, and the social environment. Reciprocally, the environment shapes and is shaped by human reality. Moral development takes place in this dynamic environment and moderates an individual's participation in, and construction of, social structure. Social and ecological degradation, as a single issue, are outcomes of the way we as individuals, as groups, and as a species, internalize or adapt and adjust social structures. An understanding of the moral development process as it relates to the process of socialization sheds light on the entire process, resulting in the degradation of Earth. Through an understanding of human morality, pedagogy for restoration begins to illuminate itself. This chapter begins by examining what may be evolutionary roots for human empathy and the potential pitfalls of such an approach and explores empathy, its possible evolutionary basis, and its limitations. Chapter 4 explores moral development and moral principles. Chapter 5 explores how and why people may act prosocially or antisocially as a result of their moral development.

· 3 ·

EMPATHY

> Empathy is a necessary prerequisite for altruism and compassion, therefore, the roots of human morality can be traced through our biology to the evolution of our species. We are, by nature, moral creatures. (Hastings, Zahn-Waxler, & McShane, 2006, p. 504)

Evolutionary Examination of Empathy and Morality

An evolutionary approach to moral development suggests that "helping genes" have been necessary for the survival of the human species (Hoffman, 1981). As social creatures, the human species' survival required that we not consider only our own self-interest in our interactions with one another. Only by working together and considering one another's needs were we able to survive the test of time (Hoffman, 2000). Humans are thinking and feeling beings "built in such a way that they can involuntarily and forcefully experience another's emotion—that their distress is often contingent not on their own but someone else's painful experience" (Hoffman, 2000, p. 5). I find it incredibly profound that the human desire for justice (as a consequence of empathy) is not rooted in an egoistic self-interest, or logical understanding of the need for reciprocity, but is an involuntary emotional component of our genetic makeup.

Social Darwinism

Before delving into a discussion of the evolutionary basis for human empathy, I want to consider the potential pitfalls of such an approach.

> The reality is that evolutionary perspectives on human behavior frequently incite controversy, even among the scientists themselves. Evolutionary theory is one of the most fertile, wide-ranging, and inspiring of all scientific ideas.... However the legitimacy of this exercise is at the center of a heated controversy that has raged for more than a century. Ultimately, the disquiet traces back to past misuses of evolutionary reasoning to bolster prejudiced ideas and ideologies.... Most researchers within the social sciences and humanities remain extremely uncomfortable with evolutionary approaches. (Laland & Brown, 2011, p. 2)

Early theories of natural selection were misinterpreted and distorted into Social Darwinism and became instrumental in the formalization of race and racial hierarchies as scientific concepts (e.g., Morton, 1839) and inspired eugenics movements in the United States as well as the Holocaust in Europe (McWorter, 2010). Although "past misuses of evolutionary reasoning" do not condemn all modern evolutionary theory, we have not escaped (scientifically or socially) the legacy of racism and eugenics. Therefore, we must be extremely cautious and critical when employing evolutionary theory.

I want to consider Darwin's own role in the distortion of the theory of natural selection. Darwin did not share significant application of natural selection to humans publicly until he had spent nearly ten years carefully considering it because he was well aware of the potential implications of doing so (Laland & Brown, 2011).

> It is largely by distorting Darwinian thinking that evolution has been used to justify prejudice and inequality. Most of the negative features sometimes unfairly attributed to evolution, including prejudice, racism, sexism, genetic determinism, and Social Darwinism, do not come from Darwin but from others who twisted his theory. (Laland & Brown, 2011, p. 47)

When Darwin did publish on the ways that natural selection acted on humans, some of those considerations of race and sex reflect prejudice by today's standards (Laland & Brown, 2011). However, Darwin explicitly rejected Social Darwinism and much of the distortions of natural selection would be unfairly attributed to him.

It was largely the work of Darwin's cousin Francis Galton who distorted natural selection into a theory of eugenics (Laland & Brown, 2011). However, the path from Darwin's theory of natural selection to atrocity would be unfairly attributed to any single individual. It was a long progression of distortions of Darwin's theory into a hierarchical conception of life in which some species are "higher" and some "lower"[1] and the formalization of races among humans that led to conceptions of racial hierarchy. This racial hierarchy was used to establish an ideology of white superiority by suggesting that Europeans had larger brains and were the most evolved race (Morton, 1839). Such racial hierarchy was used to justify the oppression of "lower" humans and, in the most extreme cases, the genocide of undesirable races.

Ultimately, this warped concept of human evolution was the ideology behind the Holocaust. However, there was an active eugenics movement in United States preceding the Nazi atrocities. The Holocaust was a result of already widely prevalent Social Darwinism and racial hierarchy (Edgcomb, 1993; McWorter, 2010). In fact, the Nazis explicitly stated that they were enacting what U.S. eugenicists were writing and mocked the United States in propaganda for calling Nazis barbaric while racism and eugenics (primarily through sterilization; however, U.S. eugenicists were considering euthanasia) were enacted in United States at the same time (Edgcomb, 1993; McWorter, 2010). Furthermore, Jewish refugees were accepted in the United States based on racial quotas. The policy was based on perceptions of the number of Jewish people who could be racially assimilated and the genetic desirability of the refugees. Those from Western Europe and accomplished academics and professionals were considered genetically superior and more assimilable (Edgcomb, 1993; McWorter, 2010).

Clearly, the misuse of evolutionary theory can result in atrocity. Modern genetics and evolutionary theory share the same roots as those from which such atrocity grew. However, I do not think this is cause to condemn all evolutionary theory and genetics or their potential. What is necessary is that we interpret meaning critically and seek knowledge for the sake of wisdom (cf. Flyvbjerg, 2001, on phronesis) rather than explicitly technological or economic progress. It isn't enough to come to empirically based judgments regarding epistemic truth. We must consider the origin of certain truths and how such truths have and can be used. We have to consider the practical application and implication of knowledge as both ethical and empirical.

The theory of evolution, properly understood, could be imagined as a branching tree in which each branch represents speciation, but not progress toward some end goal or higher species. Social Darwinism and eugenics reduce that tree to a branchless snag: "a linear, progressive concept of change perhaps inevitably engenders prejudice as some evolved forms must be regarded as more advanced, or 'higher,' than others" (Laland & Brown, 2011, p. 46).

Evolution of empathy and altruism

With the previous qualification, some understanding of what may be innately human does provide insight into the nature of empathy and human morality. However, we must constantly keep in mind that evolutionary dispositions do not dictate modern behavior in a deterministic sense. It is likely that such genetic influences do relate to our evolutionary context, but innate and acquired behavior can never be considered independent of each other. Human behavior is channeled but not predetermined by evolved predispositions (Laland & Brown, 2011).

Altruism, simply defined, is voluntary behavior that promotes the needs of another at some expense of the altruistic actor (Hastings et al., 2006) with no expectation of external reward (Oliner & Oliner, 1988). There is fossil evidence in human history to suggest that "early humans did not live alone but in small, nomadic hunting and gathering groups. Such groups produced more offspring (who presumably continued to live in groups) than those individuals not living in groups" (Hoffman, 1981, pp. 111–112). Altruism is posited to be an adaptation to group living. The Darwinian concept of survival of the fittest may appear very individualistic, placing individuals in competition with one another. This would suggest an evolutionary basis for egoistic, self-serving behavior—the opposite of altruism. The least altruistic individuals would be the most likely to reproduce and any genetic disposition toward altruism would be selected away (Hoffman, 1981). However, current understanding of evolution recognizes the role of natural selection in promoting both group and individual adaptations. In the case of avoiding predation, a member of a social group may defend the group or draw the predator away at risk to their own life (Hastings et al., 2006). Conversely, in an entirely nonaltruistic social group the individual may escape predation but reduce overall group survival. Furthermore, all "individuals benefit when the group cooperates to defend itself from predators" (Hastings et al., 2006, p. 485). Therefore natural selection would favor the social group with a predisposition for altruism and cooperation because it would increase group survival (Hastings et al., 2006).

The evolution of concern for others (empathy) may be connected to the evolution of mammals 180 million years ago. Mammals, especially, but not exclusively, nurture, nourish, and protect their young, in contrast to most reptiles who abandon their eggs shortly after laying them. Hastings et al. (2006) suggested that in these longer-term relationships between parent and offspring, the parent would be more frequently exposed to the feelings associated with pain, separation, and suffering of the offspring. Any trait that improved the ability of a parent to detect and respond to the offspring's feelings would improve the chance of survival. With this increase in survival of offspring, a predisposition for empathy as an affective and cognitive process may have been selected for. This may also be supported by evolutionary changes in the mammalian brain (Hastings et al., 2006).

Although I primarily focus on social dynamics among and between humans, I want to note that empathy is not a uniquely human characteristic. Altruistic behavior has been observed in other animal species from primates to rats. "Cooperation and helping behavior are observed in species other than humans including several types of nonhuman primates, birds, hunting dogs, dolphins, and whales" (Hastings et al., 2006, p. 485). Humans may also feel empathy or similar feelings of compassion or concern based in moral principle for non-human animals and the natural environment (Opotow, 1990; Plous, 2003).

There are many influences determining when and whether an individual will act altruistically or cooperatively. In some species, kin selection disposes individuals to favor those most genetically similar to them (Laland & Brown, 2011). Considerations of reciprocity certainly are also very influential in social species. Vampire bats, for example, will regurgitate food for other bats who return to the roost having not found enough food and are therefore at risk of starvation. Since they live in relatively stable groups, it seems that although no immediate benefit is observed, the bats have some expectation that the favor will be returned if they find themselves facing starvation. In that way, their altruism is constrained by a demand for reciprocity (Laland & Brown, 2011). Among early humans, such reciprocal altruism likely involved identifying cheaters who received benefits and did not reciprocate and may have influenced group selection. Reciprocal altruism may also be fundamental to human concepts of justice (Laland & Brown, 2011).

Simple theories of social and biological determination dictated by evolutionary predispositions toward helping one another are not able to account for the complexity of social dynamics among modern humans. Furthermore, I don't see utility in determinism, which seems to me a slippery slope to fatalism. Genetics alone rarely influence human behavior independently of our unique accumulation of life experience. Even if sociobiology and related fields eventually provide a complete understanding of human nature, there would be no reason or even possibility of submitting to it. As Freire (1998) suggested, we are conditioned but not determined; by becoming aware of the ways we are conditioned, we can transcend those conditions. Still, keeping Horkheimer and Adorno (1944/2002) in mind, it's difficult to reconcile the relationship between knowledge and domination. They suggested that to know nature is to dominate it and the same could potentially be true of human nature (Horkheimer & Adorno, 1944/2002). I remain in Freire's (1998) corner with critical hope and believe the more we understand human nature, the better equipped we will be to circumvent some aspects (e.g., selfish genes) and foster others (e.g., altruism) through socialization.

Furthermore, we should not consider elements of "human nature," such as altruism, cooperation, and morality, as static. Certainly, if these dispositions evolved they continue to evolve. Human nature two million years ago certainly is not the same as human nature today, if such a thing even exists. At best, early human history can provide clues but not answers to the timeless questions of human nature.

Bookchin (1982) suggested that civilization itself is a potentially deterministic influence on human behavior and wrote that civilization almost autonomously shapes humans and advances not with humanity, but despite us. Bookchin meant that we have established a society structured in a way that humans internalize and reconstruct social structure without consciously doing so and thus civilization is capable of reproducing itself without anyone explicitly consenting. Humans have become uniquely disconnected from the natural world and are no longer governed exclusively by the laws of nature. The governance of humans now plays a significant role, alongside the laws of nature, in shaping the future for human society and the entire planet. I hypothesize that humans can move beyond, or into alignment with, what may or may not be human nature through education, socialization, and governance.

The Process of Empathy

As discussed in the previous section, it is likely an interaction between genetic dispositions and the socialization process that determines our behavior. Regardless of the influence that genetic disposition has on behavior, educators can't do much to influence anyone's genetics, but we can certainly be influential in the socialization process. The following is an attempt to understand the processes that ultimately cause people to care about the ecological integrity of the planet and the well-being of fellow humans. With occasional digressions back into evolutionary influences, I focus on an understanding of social influences and psychological processes of empathy and moral development. I believe that this model for moral development offers significant insight for educators, but all models by definition, represent processes that are not entirely understood. The following model of moral development cannot possibly account for the development or actions of all people. It offers a unique insight along specific dimensions, but provides little or none along others.

Hoffman's (2000) theory of empathy and moral development suggests that empathy is an involuntary and emotional process that leads to feelings of distress when witnessing another's pain. Such distress encourages one to assist a person in pain and to avoid causing others pain. According to Hoffman (2000), this process is the basis for human morality. The model provides incredible insight into human behavior and I examine it very thoroughly.

Empathy is most simply described in terms of its outcome in which one experiences feelings more congruent with another's situation than with one's own (Hoffman, 2000) One outcome is state matching: feeling what another feels. However, as Garbarino (2011) suggested, empathy is never complete and someone can never know exactly what another is experiencing (cf. Decety & Jackson, 2006). State matching is an affective (feeling) process. There is also a distinctly cognitive (thinking) form of empathy: an awareness or recognition of another's internal states (Hoffman, 2000). In reality, cognitive and affective empathy are generally experienced in tandem. Strictly cognitive empathy could cause someone to empathically feel anger on behalf of another who was treated unjustly even though that person may feel pain or fear. Strictly affective empathy should dictate that the observer has similar feelings of pain and fear to the victim. In reality, the observer may empathically feel some combination of anger, fear, and pain. Hoffman (2000) emphasizes affect, but conceptualizes empathy as a process of thinking, feeling, and acting rather than a simple outcome (e.g., affective state matching). Hoffman's primary concern is empathic motivation for prosocial behavior and does not generally disambiguate empathy and sympathy. Thus, before continuing, I find it important to explicitly state the caveat that an affective response is not necessary for one to feel concern (sympathy) for another (Decety & Michalska, 2010). As explored in the following section, empathic overarousal (too much empathy) may actually reduce sympathy.

Prosocial motivation

The human capacity for empathy leads people to have feelings of empathic distress when others are hurt or mistreated (Hoffman, 2000). The long-term outcome for the observer is the formation and internalization of moral principles that function to avoid or alleviate future empathic distress and associated feelings of guilt. Hoffman (2000) described guilt as "a painful feeling of disesteem for oneself, usually accompanied by a sense of urgency, tension, and regret, that results from empathic feeling for someone in distress, combined with awareness of being the cause of that distress" (p. 114).

Guilt is a powerful motivator, but there are other possible motivators. For example, reciprocal altruism suggests that we may be motivated to help one another by the expectation that if we help someone, they will return the favor when we are in need (Laland & Brown, 2011). However, I want to consider the possibility of intrinsic motivation to help others without any associated survival strategy or feelings of guilt. I want to briefly consider joy. Guilt more significantly than joy motivates us to avoid harming one another, and the ways in which we do nice things for one another may be less empathically and morally motivated. Still, I want to take a moment to consider the importance of positive motivation. Writing of involvement in social and political activism, Berman (1997) posited that "joy comes from living in such a way as to promote good in the world, being fully present in one's efforts, and experiencing a sense of connection with others with a struggle, and with the welfare of the planet as a whole" (p. 78).

Haidt (2003) considered it "an emotion triggered by people behaving in a virtuous, pure, or superhuman way," such as helping someone in need, and referred to it as elevation. Haidt suggested that elevation is experienced affectively as warm, pleasant, tingling feelings, or openness. Upon witnessing such virtuous acts and experiencing elevation, one is uplifted and motivated to behave virtuously and help others (Haidt, 2003). Studies following Haidt's concept of elevation have supported the relationship between witnessing a good deed, experiencing elevation, and engaging in prosocial behavior (Thomson & Siegel, 2013). Garbarino (2011) similarly referred to "transformational grace," which could be described as a state of being intensely and perpetually positively motivated. I suspect that we can come to know joy and grace only through guilt and shame and that neither pair could exist in the absence of the other. However, despite the significant prosocial role of guilt, much of the contemporary work of positive psychology suggests a greater human potential through positive motivation.

Empathic arousal

Hoffman (2000) identified five ways in which empathy is aroused in humans: mimicry, classical conditioning, direct association, mediated association, and role-taking. Mimicry is a two-step process that occurs in rapid succession. Humans tend to (1) automatically imitate the facial expression, vocal expression, and posture of one another, and then (2) experience feedback based on that imitation which causes them to feel a related emotional state. Mimicry could be considered an egoistic experience of empathy because the observer's feelings result from their own imitation of someone else's expression, but not from a recognition of the other's feeling. Mimicry is considered the most basic form of empathy and has been observed in infants in their first day of life. It may also underlie all of the other ways in which empathy is aroused. A final note on mimicry—the simple act of mimicry not only arouses empathy as a prosocial motive, but may in itself be a prosocial act. Mimicry communicates awareness, involvement, connection, and understanding of another's emotional state and even solidarity between the observer and the other (Hoffman, 2000).

Classical conditioning can also be interpreted in much the same way as mimicry, as a two-step process occurring in rapid succession. However, rather than the first step being imitation, the observer associates the observed situation or display of emotion with emotions that they have previously experienced. The second step is a similar process of feedback, which is again quite egoistic, wherein the observer experiences the emotion that they associated with that observation. The conditioned emotion isn't necessarily the same emotion that the observed other experienced. Classical conditioning has been observed in one-day-olds when a sucking response was conditioned to stroking the newborn's forehead (Hoffman, 2000).

Direct association is an empathic response to a situation wherein the observer has had a similar experience. The observer associates the similar experience directly with the situation that they are observing and feels empathy related to recollection of the similar experience (Hoffman, 2000). For example, a child could feel empathic pain in response to seeing their parent accidentally hurt, because of the child's recollection of the pain of having been hurt in the past. In the same example, mimicry might be at work if the child winced in response to the parent's facial and vocal expression of pain and experienced empathic distress as a direct result of the child's own expression of pain. Classical conditioning might be at work if the child became afraid because the parent frequently became angry after being injured. These three modes of empathic distress are considered to be automatic, quick acting, and involuntary (Hoffman, 2000). The preceding affective forms of empathy are the earliest to "come online" in the course of infancy and early childhood development (Decety & Michalska, 2010).

Mediated association generally requires language and is a more developmentally complex process (Hoffman, 2000). Most frequently, the empathic feelings are moderated verbally. For example, someone could arouse an empathic response by saying, "I am sad." The observer would process this information and might feel empathic distress that would likely involve some component of the other modes of empathic arousal. Mediated association can also be cognitively mediated. Hoffman (2000) gives an example of someone who regularly spoke to their grandmother who was in a coma. The grandchild felt empathy based on their previous relationship with the grandmother, knowledge of the patient's situation, and the assumption that the grandmother could hear them, while receiving no cues indicating their grandmother's internal state. It becomes clear that mediated association requires a certain level of abstraction and theory of mind (conception of others as distinctly unique from the self with its own internal states and goals). Theory of mind is not necessarily prerequisite to the first three modes of empathic arousal, which could conceivably be entirely egoistic. Most neurotypical adults have a theory of mind. Other animals, such as chimps, also have a theory of mind (Call & Tomasello, 2008). Interestingly, much of the evidence that supports Call and Tomasello's conclusion involves chimps attributing theory of mind to humans. However, as humans, for more than thirty years we have hotly debated whether chimps have a theory of mind (Call & Tomasello, 2008).

The final mode of empathic arousal that Hoffman (2000) discussed is role-taking (perspective taking). This is the act of cognitively putting oneself in another's place and imagining how the other might feel. Role-taking is developmentally advanced, and the ways in which an observer practices role-taking is reflective of that individual's personal development. Role-taking may be either self-focused or other-focused. Self-focused role-taking is the act of observing someone else and imagining how one would feel in that person's situation. It is possible, but not ensured, that this will lead to the observer having empathic feelings that are similar to the other's. Other-focused role-taking is the act of observing someone else and imagining how that person feels. This requires the observer to have the ability to conceptualize others as separate and unique individuals. Other-focused role-taking can be enhanced by observers' knowledge about those they observe.

For example, if someone witnessed a mugging, self-focused role-taking could lead the observer to feel angry while the victim feels unsafe and frightened. It is easy to imagine how this could lead to victim blaming, especially if the observer was an able-bodied master of karate who would feel safe confronting the mugger. If the victim was a less able-bodied or elderly individual, other-focused role-taking might result in the observer feeling a more empathic understanding of the victim's fear. How and why folks act on empathic distress follows, but in the former case, if the master of karate chose to help, a self-focused role-taking might lead to helping by confronting the mugger, and other-focused role-taking might be more likely to lead the observer to focus on comforting or protecting the victim.

Self-focused and other-focused role-taking can both be prosocial motives, and neither necessarily results in the empathic observer having feelings similar to the one whom they observe. While other-focused role-taking is a more cognitively advanced task, both can serve as prosocial motives. Commonly, role-taking is a combination of self- and other-focused. Additionally, all of the modes of empathic arousal complement one another and may be experienced in various combinations. In the mugging example, it is likely that the observer would feel empathy not only based on role-taking, but also by observation of the victim's reaction in the form of mimicry, classical conditioning, direct association, or the victim's verbal response through mediated association.

The relationship among ecological degradation, mediated association, and perspective taking warrant further exploration. Plous (2003) observed humans feeling empathy for other animals depicted being beaten. Franklin et al. (2013) provided neurological support for Plous's (2003) claim that humans feel empathy for non-human animals and that many of the same regions of the brain associated with empathy become active whether observing a human or other animal suffering. However, Franklin et al. (2013) noted that the processes are not a match and that different regions of the brain are significantly more active when a person is observing human suffering. Franklin et al. (2013) suspected that this difference may be related to the way in which humans perceive animals differently, which may relate to cultural and individual differences. It seems conceivable that some humans would directly associate a non-human animal's pain with their own experiences of pain. However, knowledge or beliefs about other animal species (for example, beliefs about similarity to non-human animals) probably do cognitively mediate the empathy that we feel for them (empathic bias is explored in the following section). Thus, by encouraging people to take the perspective of a non-human animal, they feel greater empathy and thus concern for the animal (Berenguer, 2010; Schultz, 2000).

In the case of application of empathic arousal to broader environmental issues, certain environmental (social and ecological) knowledge related to the consequences of our actions likely mediates our empathic arousal regarding environmental issues (Berenguer, 2010). Furthermore, although the welfare of animals is intrinsically important, concern for an individual non-human animal has been shown to increase concern, more generally, for the environment (Myers, Saunders, & Garrett, 2003, 2004). Thus, encouraging empathic concern for animals is important not only for animals as individuals and as a species, but also for ecological restoration as a whole. I revisit concern for other animals in the Moral Exclusion section in Chapter 5.

Discipline and intervention

Hoffman (2000) suggested that feelings of empathic distress motivate prosocial or helping behavior and moral internalization through a translation from empathy to sympathy and associated feelings of guilt. The modes of empathic arousal that have so far been discussed primarily assume that the empathic observer is an innocent bystander and has not been involved in the transgression that resulted in the emotional state of the target. Hoffman suggests that children develop morally through their own transgressions and resulting disciplinary action. Based on those experiences, children begin to make inductions to more general situations and self-discipline through guilt. Moral development through induction builds capacity for feelings of guilt over inaction or anticipatory guilt over inaction that encourage helping behavior in situations in which the actor is an innocent bystander. It also fosters a conception of one's role as a virtual transgressor in situations in which one may feel complicit in a transgression in which a person is not directly involved.

Socialization is the key to understanding moral internalization. Hoffman (2000) specifically investigates the role of power in discipline/intervention, most significantly by children's parents, but not exclusively. The way that children are disciplined in situations in which they are the transgressor hugely influences how and when they will feel guilt and how helping behaviors will be triggered or catalyzed in the future. An understanding of this aspect of the socialization process in moral internalization will set the stage for understanding the myriad forms of guilt that humans feel, empathic biases, moral exclusion, and the moral principles that guide our interactions with one another.

Hoffman (2000) defines three forms of discipline: power assertion, love withdrawal, and induction. Power assertion can be physical, as punitive or constraining discipline, or non-physical. Power-assertive methods of discipline may be softened by explanations but all rely primarily on an assertion of power of a guardian or other authority figure over the child. An example of a punitive-physical power assertion is spanking a child and an example of constraining is picking up and moving the child. An example of a non-physical power assertion is a firm and authoritative correction such as, "No! Don't ever do that again." While this is the least conducive method to empathy development, Hoffman (2000) points out that unqualified power assertion is necessary in emergencies. Power assertion can also be used in conjunction with other discipline methods. Love withdrawal may be silent or explicit. An example of silent love withdrawal is ignoring a child who is misbehaving and an example of explicit love withdrawal is sending a child to their room, or saying, "I don't like you when you act that way." While power assertion is more likely to result in a child feeling fear, love withdrawal is more likely to result in anxiety.

Inductive interventions are most conducive to empathic and moral development. In inductive intervention, the child's guardian or other authority figure helps a child who has acted as transgressor to understand how their actions affect others. An example is a guardian telling a child, "When you hit them, it hurt them and made them cry. You should never hit someone because it hurts them. If you said you were sorry, they would feel better." Developmentally appropriate inductive interventions accomplish two things that the other forms of discipline do not. First, they call attention to the emotional distress of others, triggering the child's own empathy and fostering the child's empathic distress. Second, they help children understand their role in causing the emotional distress of others and help them to feel empathy-based guilt. This encourages children to avoid the transgression in the future and to make reparations when they do transgress. This establishes an emotional guilt-script that children eventually internalize and follow without the guidance of a guardian or authority figure: Transgression ⟶ Induction ⟶ Empathic Distress and Guilt ⟶ Reparation.

As with the modes of empathic arousal, methods of discipline rarely fall neatly into just one category. Often, discipline involves components of all three. I recall an incidence of bullying on a nature hike that I led. One of the kids had been consistently bullying the others and at one point the kid picked up a stick and chased another down the trail with it. The immediate need was to stop the dangerous situation. I've rarely raised my voice to children, but I yelled for them to stop and put down the stick. They stopped near me and started arguing. I said that I didn't care who started it but it needed to stop. Once it seemed as though the immediate danger was diminished, we discussed what had happened. Both kids were still angry and I was also upset. Our attempt to discuss the situation was not very productive. I had the entire group hike silently for the next ten minutes or so and asked them to think over what had happened. This also gave me a chance to calm down and think about the situation. When we stopped, we gathered in a circle and I explained that I hadn't meant to yell at them, but that I was scared that someone was likely to have gotten hurt. As a group, we discussed what had happened and brainstormed how to get along better and resolve disagreements as a group for the rest of our time together. At the time, I wasn't thinking about different methods of discipline or how they relate to moral development. I was just reacting. I share this story because I think it illustrates the types of discipline and their relationships in a real situation. My initial reaction was power assertive; then I attempted induction, which was unsuccessful because tensions were too high, so I used something on the order of love withdrawal by asking them to hike silently for ten minutes. After that, I tried to soften this love withdrawal with an explanation and apology and then tried induction for the second time, which seemed more successful than my first attempt. I don't know whether I handled the situation well and I often worry that the kid who had chased the other with the stick will wind up in trouble in the future. However, I've better understood the concepts of moral development through consideration of my own life. I doubt that one experience on a class field trip significantly affected the kid's established behavior and bullying often correlates with anti-social behavior in adults and the consequences are much more significant.

Guilt

With an understanding of the complexities and abstractions involved in guilt and the related empathic biases, the ways in which we come to establish our moral principles and specifically our concepts of justice become apparent.

Hoffman (2000) identified the prerequisites for guilt as: an awareness of another's distress, a sense of control over one's actions, an awareness of the consequences of one's actions, the ability to attribute causality to those actions, and the ability to tell the difference between intentional actions and accidents. This begins to extend the concept of guilt beyond guilt associated with transgression (having directly caused the pain of another) and anticipatory guilt (relating to guilt felt prior to or in contemplation of a potential transgression), and guilt over inaction (guilt felt for not helping in a bystander context) to guilt in a more abstract sense, such as guilt over virtual transgressions. Virtual transgressions are transgressions in which a transgression isn't directly committed; however, the transgressor (reasonably or unreasonably) perceives that their behavior caused harm.

Relationship guilt occurs in close personal relationships (Hoffman, 2000). It can result from transgressions ranging from forgotten appointments to betrayals of confidence, or it can relate to feeling guilty for another's distress when there is no indication that it is the result of one's own transgression. This is commonly, but not exclusively, felt in romantic relationships when people feel that their partner is upset and assume, even when told otherwise, that they must have done something wrong and they feel guilty for it. Similarly, responsibility guilt is related to feelings of responsibility for another's well-being. Responsibility guilt doesn't necessitate a close relationship. For example, in a disaster, a city official might feel responsibility guilt for the suffering of folks in that city, despite evidence that there was nothing the official could have done to prevent it (Hoffman, 2000).

Developmental guilt includes guilt related to separation from friends and family, guilt over achievement, and guilt over affluence. Hoffman (2000) stated that

> an individualistic, competitive society like ours can make growing up and pursuing normal personal goals and interests the context for virtual transgressions involving "developmental guilt." (p. 182)

It is important to understand that virtual transgressions result in real guilt and are only virtual in the sense that they are not perpetrated with an intention or awareness of their potential for harm (Hoffman, 2000). I think virtual transgressions are easiest to understand in the context of separation guilt. When a young person leaves home for the first time to attend college and leaves behind friends and family, that person may experience a feeling of guilt. This is a virtual transgression, but there is a very real chance that friends and family will miss the person and suffer emotional distress. Assuming that the student is not attending college out of malice or contempt for friends and family, this would not be considered a transgression in the dominant U.S. culture. It is considered a developmentally normal process for young people to leave home to attend college, so it is a virtual transgression.

Guilt over achievement is incrementally more complex with regard to the virtual nature of virtual transgressions (Hoffman, 2000). Rooted in social comparison, achievement is commonly a competitive endeavor. In school, kids may feel guilty for earning high grades while other students are less academically successful because they recognize the effect of social comparison on a peer's self-esteem. In other cases, achievement may be earned at the direct expense of another. Receiving a top scholarship, winning first place in a competition, or earning a promotion all dictate that someone else didn't get the scholarship, win, or get the promotion.

Guilt over affluence requires some form of exposure to folks who are less fortunate and cultural context that does not provide justification for disparities in wealth and racial superiority. Guilt over affluence is guilt felt for having more than others. While transgression guilt may be alleviated by a single act of restitution, those who feel guilt over affluence may feel compelled to participate in sustained activity to relieve human suffering (Hoffman, 2000).

> As with bystander guilt, if one does nothing one continues to feel guilty; blames victims; cognitively restructures to justify inaction, deny or justify ones affluence: "They have their pleasures and enjoy the way they live"; "I worked hard for what I have." (Hoffman, 2000, p. 186)

Guilt over affluence begins to push the limits of the virtual nature of virtual transgressions—at least to those with a socialist perspective. Many transgressions only remain virtual, in a capitalist system, with norms of meritocracy, and with the contribution of ignorance, moral bias, and moral exclusion. Without such defenses, affluence and the following transgressions are not virtual because they are at the direct expense of other people. Guilt by association refers to guilt caused by association with a particular social group that one considers culpable. In the case of affluence, it may be class association. Guilt by association could also relate to association with a political or religious group. More generally, guilt over affluence and guilt by association are guilt felt over relative advantage (Hoffman, 2000). This might more aptly be referred to as privilege guilt in a social justice context.

A final consideration, as alluded to in the description of guilt over affluence, is victim blaming. Victim blaming is a process of justifying the suffering of others. Especially in a bystander situation, victim blaming "puts psychological distance between bystander and victim and reduces the bystander's empathic distress and motivation to help" (Hoffman, 2000, p. 94). This is especially common when the bystander doesn't feel able to help.

> What may transform empathic distress into guilt feeling, is that one cannot justify and therefore does not deserve the advantages one has over the victim; that one's advantage, surviving or being affluent while others starve or die, violates the principle of fairness, justice, or reciprocity. One has an empathic feeling of injustice that is transformed into guilt feeling, because one is the beneficiary of the injustice. (Hoffman, 2000, p. 190)

An illustration of the role of guilt is provided by one of the Holocaust rescuers interviewed by Oliner and Oliner (1988):

> It was unfair that I was safe simply because I was born a Protestant. That was the main reason for me. What I did was a question of justice. It was a humble thing because I was in a privileged situation compared with other people who didn't deserve their situation at all. (p. 166)

In summary, there are four categories of guilt-arousing processes, according to Hoffman (2000): (1) guilt resulting from external intervention such as induction with children, (2) guilt spontaneously arising (as a result of internalization) following a perceived transgression, (3) guilt spontaneously following an ambiguous situation such as relationship guilt, (4) guilt following inaction as a bystander. Finally, "empathy-based transgression guilt may be the prototype for all empathy-based guilt, real or imagined" (Hoffman, 2000, p. 193). According to our feelings of guilt and the inductions that we make in response, we begin to construct moral frameworks regarding appropriate ways of interacting with one another.

Limitations of Empathy

Hoffman's (2000) examination of the limitations of empathy is strictly in consideration of empathy as a prosocial motive. Empathy is addressed as a process including empathy and sympathy rather than the simple outcome of emotional state matching. Hoffman (2000) did explicitly refer to humans as highly or less empathic, in reference to capacity for an empathic outcome such as feeling what another feels. However, the value of understanding the limitations of empathy is in understanding empathy as a process in which the most important outcome is not state matching but motivation to help. Although the intensity of emotion experienced by an empathic witness may be a significant motivator, those who Hoffman (2000) suggested are less empathic may be motivated cognitively. Thus the focus on empathic limitations is focused on ways in which (cognitive and affective) empathy may preclude helping behavior.

The limitations of empathy include potentially self-destructive and certainly limiting reactions to empathic distress, differences in the character of the empathic individual, and empathic biases that may interfere with the guilt-scripts that motivate us to help each other. The limiting reactions to empathic distress include empathic over-arousal, compassion fatigue, individual character differences, and empathic bias.

Empathic over-arousal and compassion fatigue

Empathic over-arousal is an "involuntary process that occurs when [an] observer's empathic distress becomes so painful and intolerable that it is transformed into intense feelings of personal distress, which may move the person out of the empathic mode entirely" (Hoffman, 2000, p. 198). Instances of empathic over-arousal may also result in egoistic drift, wherein the empathic distress triggers a process in which someone gets so caught up in their own feelings that they are forced to attend to their personal distress and disengage from the feelings associated with the person experiencing distress. However, empathic over-arousal can also serve as a prosocial motive by intensifying a person's attention and helping behavior, even to the point at which it becomes all-consuming if the situation does not become so overwhelming that the person disengages entirely (Hoffman, 2000).

Compassion fatigue results from prolonged or continuous states at or near the point of empathic over-arousal. These processes may also result in "vicarious traumatization" (trauma and empathic disorders are briefly discussed following the discussion of limitations to empathy). One way that folks defensively cope with or avoid compassion fatigue results from habituation. Continued exposure to another's distress over time may result in a process of habituation, which diminishes empathic distress to the point at which a person may become indifferent to another's suffering (Hoffman,

2000). Hoffman primarily used examples of medical professionals to illustrate these concepts, such as a medical student unable to cope with regular interactions with a terminally ill patient and a doctor who seems unaffected. However, compassion fatigue, habituation, and trauma also relate to situations in which youth are regularly experiencing violence (see the section "Outlying Behavior" in Chapter 5) (Garbarino, 1996). Berman (1997) and Oliner and Oliner (1988) both discussed the importance of community in overcoming compassion fatigue among contemporary U.S. activists by the former and Holocaust rescuers by the latter. Many of the folks, who continually placed themselves in situations arousing severe empathic distress, discussed their reliance on a supportive community (Berman, 1997; Oliner & Oliner, 1988).

Character difference

There are also individual characteristics that affect how an individual might respond to the emotional distress of another. The first is described as empathic tendency and suggests that some folks are more empathic than others (Hoffman, 2000). This refers to the empathic outcome of emotional state matching. Those who are considered highly empathic, in some cases, may be less able to help owing to their empathic distress (Hoffman, 2000). Those who are better equipped to regulate their emotions, handle anxiety, and cope with empathic distress are better able to avoid empathic over-arousal (Hoffman, 1981). One method of regulating emotions is distancing oneself from a patient or victim (Hoffman, 2000). By recognizing the other as distinct from the self and "maintaining a sense of whose feelings belong to whom" (Decety & Jackson, 2006, p. 56), empathic over-arousal may be reduced or avoided. The process of distancing may result in more pronounced dissociation. Oliner and Oliner (1988) found feelings of dissociation from Holocaust victims to be equally common among rescuers and non-rescuers. Finally, a sense of self-efficacy also contributed to helping behavior and reduced the risk of empathic over-arousal (Hoffman, 2000). Those who feel more able to help are more likely to help and feel less overwhelmed by the empathy that they feel. This is easy to understand in the context of guilt scripts and in the identification of empathy as a prosocial motive. Empathic distress is relieved by helping. However, helplessness and hopelessness can be exacerbated by the described character differences in increased empathic bias.

Empathic bias

Empathic bias may be related to human nature but is also deeply related to socialization. Empathic bias falls into two categories, familiarity bias and here-and-now bias. Familiarity bias refers to an empathic bias toward those with whom one is familiar, including bias that favors one's in-group, friends, and those found to be similar to oneself. The evolutionary argument for familiarity bias is that humans evolved in small groups where resources were scarce and the small groups of early humans were in competition for resources, so an empathic bias toward one's own in-group favored group survival (Hoffman, 2000).

Hoffman (2000) cites in-group bias as a factor in violent crimes in which perpetrators feel diminished transgression guilt by distancing themselves from their victims by derogating them or even viewing them as subhumans. This was certainly the case during the Holocaust (Oliner & Oliner, 1988). The nature of the relationship between friends fosters greater understanding of one another and a stronger sense of preference for one another and is referred to as friendship bias. It remains unclear, however, in Hoffman's (2000) work, whether the actual feelings of empathy are stronger between friends, or if the bias is due to the voluntary and reciprocal nature of friendship. Still, the bias toward prosocial behavior between those with whom we have close personal relationships (e.g., friends and family) is evident.

Similarity bias can also be considered more generally. Real or perceived similarity (with examples in gender and personality) results in greater expressions of empathy (Hoffman, 2000). Oliner and Oliner (1988) found that many rescuers seemed to assume Jewish people were similar to them, regardless of any experience with them. Actual familiarity with Jewish people as measured by rescuer accounts of social relations with Jewish people and knowledge of Jewish culture was not found to be an indicator either for or against rescue. I posit that an observer must feel some sense of connection to the target in order to feel empathy. A perception that the target is similar to oneself (even in the vaguest way) establishes some affinity with (if not for, as in friendship bias) the target. Such an association may be necessary for empathy. Increased familiarity with the target may further increase the empathic connection, especially if the observer is fond of the target. Similarity bias has also been shown to affect human empathy for non-human animals. We feel greater empathy for animals that we perceive to be more similar to us, such as monkeys, than animals that we perceive to be less similar, such as pheasants (Plous, 2003).

Empathic bias is likely a significant factor in racism. In non-human species, kin-selection suggests a bias toward the most genetically similar members of a species (Laland & Brown, 2011). However, in humans it has been well proved that similarities in phenotypic variation such as skin color do not correlate with genetic similarity (Guest, 2014). Humans have never become genetically different enough from other contemporaneously living humans to prevent interbreeding;[2] thus, significant genetic drift among any sub-population of humans is extremely unlikely (Guest, 2014). *Homo sapiens* have always shared the same genetic pool and genetic variation has occurred in geographic clines (Guest, 2014). Archaic *Homo sapiens* (Neanderthals) are believed to have coexisted with modern *Homo sapiens* for around 4,000 years from 32,000 years before present (yBP) to 28,000 yBP, and a consensus on the cause of the disappearance of Neanderthals has not been reached (Guest, 2014). However, there is evidence that Neanderthals and modern *Homo sapiens* did interbreed and Neanderthal genes may be present in the modern human gene pool (Guest, 2014).

There may have never, at any point in human evolution, been a genetic basis for multiple races among humans (if the inability to interbreed defines speciation). Evolved bias toward one's own family and friends is reasonable, but an evolved preference for a larger ethnic or racial group, beyond a concept of an extended family, must be primarily socially influenced. None of this denies the very real existence of familiarity bias, but if causality for a racial in-group bias was attributed to genetics rather than socialization, it would imply that there is some evolutionary basis for race. Race is entirely socially constructed, but empathic bias based on familiarity may be biological. If through socialization we are racialized, then racial in-group bias becomes real even if there is no genetic precedent for it. Simply stated: "Situations defined as real are real in their consequences" (Oliner & Oliner, 1988, p. 260).

We may feel more similar to others based on familiarity or in-group membership and thus be disposed to empathic bias. For humans to have a sense of familiarity based on likely common experiences related to the process of racialization seems reasonable. At the individual level, and when not taken to the extreme, this is not necessarily problematic and can be quite healthy (e.g., affinity groups). Obviously, this becomes problematic when one person views another as so dissimilar that they fail to feel any sympathy (see the section "Moral Exclusion" in Chapter 5). This becomes extremely problematic when a powerful group acts drastically according to its members' own in-group bias and systematically denies justice to those outside their group.

The final type of empathic bias is here-and-now bias and suggests that humans have heightened receptivity to the events that are spatially and temporally immediate (Hoffman, 2001). This suggests that people would feel a greater sense of empathy in a situation in which they are bystanders or transgressors than in virtual transgressions. During the Holocaust, 70% of the rescuers made their decision to help within minutes of consideration and more than half were motivated to help by the immediacy of the violation of the social norms that they valued (Oliner & Oliner, 1988). Furthermore, 78% of rescuers took their first action because someone asked them for help, making the situation real and immediate (Oliner & Oliner, 1988).

Although this chapter is dedicated to a descriptive investigation of the empathic process, I refuse to let this section on the limitations of empathy stand as an account of human deficits. Regardless of whether empathic limitations are part of human nature, we are not bound by them. The limitations to human empathy present an educational opportunity for us to strive with our students to reach a greater potential. It is clear, even before delving into human morality and the moral development process, that fostering and facilitating empathic induction is a significant point of behavioral intervention. We can facilitate experiences to build self-efficacy and teach coping skills to encourage sympathetic action and reduce empathic over-arousal. We can help students to see the connections between their local actions and global consequences and the interconnectedness of the world to reduce here-and-now bias. We can also support students in exploring their own identities and honoring world cultures with cross-cultural curriculum and encourage an appreciation of the similarities and differences among all peoples to encourage empathy by positively using similarity and familiarity biases (see Chapter 7, "Pedagogy of Identity"). Finally, the limitations to empathy are functionally necessary to a certain extent. We simply don't have the capacity mentally, emotionally, or practically to feel empathy and sympathy and to take restitutionary action in response to every transgression and injustice in the world. As Hoffman (2000) suggested, "If people empathized with everyone in distress and tried to help them all equally, society might quickly come to a halt" (p. 14).

· 4 ·

MORALITY

The development of empathy and its relation to guilt, in consideration of empathic bias, and contemporaneous cognitive development shape one's moral development (Hoffman, 2000). According to one's experience, throughout the empathy and moral development process one's sense of morality may differ significantly from others. Thus, it is useful to disambiguate morality into distinctly philosophical and psychological components. The psychological component relates to how one's sense of morality develops and the philosophical component relates to the nature of morality (Kahn, 1999). Hoffman (2000) generally proceeded without such disambiguation and favors a psychological approach to morality focused on development. According to Hoffman (2000), one's philosophic concept of morality is significantly a result of socialization in the moral development process (following David Hume).[1]

I proceed with Hoffman's (2000) framework for morality, which provides significant depth and insight, but it will be useful to begin with a brief exploration of morality as a more distinctly deontological idea. Kahn (1999) considered morality more philosophically than did Hoffman (2000) and suggested that morality is objective (Immanuel Kant as interpreted by Eliot Turiel). Kahn (1999) defined two categories of moral judgments: obligatory moral judgments and discretionary moral judgments. Obligatory judgments are defined as (1) universally applicable to all morally similar situations; (2) not contingent on societal rules, laws, or conventions; and (3) justified by principles such as fairness, justice, welfare, or rights. However, Kahn (1999) did not ignore cultural context in the universality of obligatory moral judgments as illuminated by the following example: devout Hindus believe that it is immoral for a widow to eat fish. At first glance, there seems to be nothing universal about this, but in the context of Hindu culture, if the widow were to eat fish it would torment her husband's spirit and cause the widow to suffer. Kahn (1999) pointed out that these beliefs may differ from our own personal beliefs but that the "underlying concern for the welfare of others is congruent with our own" (p. 204). Kahn's (1999) pluralistic considerations are very strictly limited by the concept of moral universality, though.

The second qualification that obligatory moral judgments are not contingent on societal rules, laws, or conventions relates to the universality of the principles. Children seem to easily conceptualize this. When presented with a scenario in which the principal of a school has announced that it is OK to hit other students, and asked whether it would be OK to hit another student, most children will respond that would still not be OK because it is never OK to hit someone. However, if presented with a scenario wherein the principal has declared that is OK to call teachers by their first name rather than Mr. or Mrs. and asked whether that is OK, most children respond that it would be OK as long as it is OK with the teacher. The distinction is that the first case violates a moral principle that most children learn and internalize early in life and the second is merely a social convention and that violating the first causes harm whereas violating the second does not cause harm. The former is considered the moral domain and the latter is the conventional domain.

Social conventions and moral norms are culturally constructed and may serve to maintain order and structure to social interactions and thus may be prosocial. For that reason, following norms and conventions is often a moral judgment even though the norms or conventions are not morally obligatory. However, moral norms are not necessarily prosocial or aligned with one's individual moral judgments, which may lead to significant conflict. Thus, it may be very difficult to determine which domain a conflict resides in and may not be as objective as Kahn (1999) and the social domain theorists suggest, wherein the moral domain always trumps the conventional. If morality is relative, then what is posited to be morally universal is actually relative to the status quo and obligatory moral judgments would largely reduce to a matter of conformity. I explore moral relativism in greater detail in Chapter 5.

Discretionary moral judgments are beyond what is obligatory or what is demanded by a situation. They may be judgments that would encourage morally "good" (or supererogatory) actions that are not morally mandated (obligatory judgments). Thus, inaction wouldn't be considered immoral. One example of a discretionary moral judgment that Kahn (1999) gives is the decision whether or not to make a charitable donation. Not donating to charity isn't generally a moral transgression (in Western culture), but many would suggest that it *is* a good thing to do.

Kahn (1999) also categorized moral judgments into positive and negative morals (i.e., dos and don'ts). Kahn (1999) found positive-discretionary moral action to be the most praiseworthy, followed by positive-obligatory moral action, and negative obligatory moral action as least praiseworthy. The interpretation that Kahn (1999) focused on is that the least praiseworthy actions are so stringently applied that they are not praiseworthy. Thus, we place greater emphasis on negative morals (Kahn, 1999). This seems consistent with Hoffman's (2000) emphasis on guilt as a motivator because it primarily motivates folks to avoid transgressions. It also reflects an orientation toward action over inaction (e.g., picking up litter over not littering). For example, action in accordance with negative-obligatory morals, such as not littering doesn't win praise, while violation of a negative-obligatory moral (littering) brings scorn. In contradistinction, action in accordance with positive-discretionary morals, such as picking up litter is praiseworthy, while inaction (not picking up every piece of litter we pass) usually warrants neither praise nor scorn.

In consideration of the reality of degradation in the world, Kahn's (1999) presumption that morality is universal led to its logical conclusion: "it is less the case that societies differ morally, and more the case that some societies (ours included) are involved explicitly in immoral practices" (pp. 205–206). I refuse such a deficit approach and find it imperative that this be reframed. Kahn's position is a rational conclusion given the assumption of obligatory moral judgments that are always entirely salient. Rather than believing in "iniquity that smites the land,"[2] I follow Hoffman (2000) in more nuanced exploration of moral development. As Hoffman (2000) explained, in the course of moral development, differing experiences (and character differences) may cause people to recognize and respond to moral conflicts differently and with greater or lesser influence on different moral principles. Hoffman's (2000) approach leaves more room for a sympathetic perspective on human behavior and seems less conducive to such broad condemnation of societies as explicitly immoral. However, bear in mind that while Hoffman (2000) attempted to understand variability in moral judgments, Hoffman (2000) also rejected moral relativism.

Moral Development

The moral development process that Hoffman's (2000) theory modeled is very similar to the empathy development process. Hoffman's theory assumes affective primacy and focuses most significantly on socialization but does not ignore the influence of cognitive development as explored by Piaget and Kohlberg. In this section, I center on Hoffman's moral development model while attempting to consider, critique, and integrate multiple moral development models. Moral development begins with personal experiences as a bystander or victim and socialization—primarily in the home through induction guided by a parent or guardian (Hoffman, 2000). The induction step continues with socialization beyond the home: in peer interactions, in school, and by media. Thus begins a process of abstraction and organization of moral principles based on those experiences. Finally, a phase of commitment takes place, during which the moral principles are accepted and internalized. Beyond that point, triggering events may make the principles more concrete and dictate moral action or restructuring of moral principles. The nature of triggering or catalyzing events is examined at the end of Chapter 5.

The foundation of moral development as conceptualized by Hoffman (2000) is empathy development. The capacity for (or, more precisely, tendency toward) empathy develops in accordance with the brain's development. Infants experience empathy globally and do not distinguish others' emotions from their own. For example, the newborn reactive cry is a tendency for newborn infants to cry when they hear another infant cry (Hoffman, 2000). "Indistinguishable from the spontaneous cry of an infant who is in actual discomfort," it is "an innate, isomorphic response to a cry of another being of the same species" (Hoffman, 2000, p. 65). Later in early childhood, children become aware of another's distress but they respond as if it were their own. Thus, their response to another's distress is to seek comfort for themselves. Next, children's experience with empathy becomes quasi egocentric (generally early in their second year). In this stage, children are able to distinguish another's distress from their own; however, children still have difficulty conceptualizing others as having unique inner states. Thus, a prosocially motivated quasi-egocentric child would attempt to comfort another child by the same means that they would comfort themself. Finally, by late in the second year children begin to understand others as unique individuals with distinct needs and motivations and begin to more significantly practice more developed forms of empathy (i.e., mediated association and other-focused role-taking). Later, and into adulthood, children begin considering not only the immediate distress of an individual, but also that person's life condition (Hoffman, 2000).

An early concept of right and wrong that precedes significant empathic and moral development is deference to authority (Gibbs, 2014). Children in the egocentric stage of empathy development may not significantly consider how their actions affect others, but will likely be aware of rules and the consequences for breaking those rules (e.g., never hit, because you will get in trouble if you do hit). The moral development process most significantly begins when the child starts to make inductions in accordance with empathic distress or guilt experienced as observer or transgressor. These inductions lead young children to generalized conclusions about the consequences of their behavior and ultimately, based on those consequences, children begin to formulate concepts of right and wrong. For young children, empathic induction likely addresses the most immediately salient elements of the transgression and the target's response (e.g., never steal sister's toy because she will cry). As children cognitively develop, their induction process becomes more encompassing and the inductions are integrated into a more comprehensive moral framework. As children enter school, their social environments and influences begin to expand. Children continue making inductions, achieve a greater social understanding beyond the home, and their perspectives become less egocentric. This process is the result of cognitive and affective development as well as socialization (Gibbs, 2014; Hoffman, 2000).

The abstraction and organization stage involves peer interactions during which children begin to accommodate other perspectives and claims (Hoffman, 2001). This decentering and demand for social reciprocity enables children to understand bias in their own and others' claims, and induction helps them become receptive to other claims and form empathy-based concepts of fairness (Hoffman, 2001). However, theorists such as Kohlberg suggested the cognitive development process leads to universal tendencies in human cognition, and in turn morality (Gibbs, 2014). Gibbs (2014) seeks a middle ground and argues for affective-cognitive co-primacy. Following Gibbs (2014), I favor co-primacy. Depending on one's own experiences and cultural context different emphasis on various moral principles may take precedent

(Hoffman, 2000). However, cognitive development likely increases a child's tendency[3] toward greater social understanding (Gibbs, 2014). Furthermore there may be elements of morality that universally result from the way our brains develop and function (i.e., objective morality). The most important conclusion for my purposes is that all but the most extreme of outliers (see the section "Outlying Behavior" in Chapter 5) develop a moral framework through some process of abstraction and organization.

The demand for reciprocity among humans may be a universal process. Although interpreted and emphasized differently among individuals and cultures, reciprocity seems to be universally considered among humans (Gibbs, 2014). It is possible that reciprocity is a social norm that is widely (or universally) present in human societies for its "functional value for promoting and stabilizing social relationships" (Gibbs, 2014, p. 56). It may also be that reciprocity is universally a result of cognitive construction and "a natural preference for balance and harmony" (Gibbs, 2014, p. 56). I favor Gibbs's (2014) conclusion that reciprocity is both cognitive-developmental and normative. An affective response to violations of reciprocity and motivation to restore harmony may not be significantly influenced by the domain in which reciprocity resides. Furthermore, it seems reasonable to expect that regardless of degradation of society some social norms must be aligned with the moral judgments of the majority of individuals. After all, social norms are socially constructed. Regardless of the subjectivity implicated by a moral model that maintains the primacy of socialization, there must be a universal need for some sort of fairness or harmony among (but not exclusively among) humans.

On the point of reciprocity and subjectivity, the cognitive-developmental theory warrants more attention. As one advances through Kohlberg's moral stages, one's concept of reciprocity is expected to change. Hoffman (2000) saw reciprocity as neutral and with the potential for prosocial or non-prosocial motivation. However, based on Kohlberg's theory, Gibbs (2014) suggested that reciprocity becomes distinctly and universally prosocial because the emphasis shifts from the "immature" tit for tat mentality to the "mature" consideration of how one prefers to be treated in one's treatment of others. Essentially, the "mature" form of reciprocity is a merged form of reciprocity and perspective-taking without a strict (or vengeful) demand for reciprocation from others. This advancement through the stages is characterized by a transition starting with deference to authority, to the recognition of multiple perspectives (relativism), and then to a universal moral framework.

Kohlberg cunningly established a universal invariant sequence of moral development that suggested a universal concept of morality was a result of advancing through a hierarchical concept of morality. Thus,

> unbridled relativism seemed to Kohlberg to be inherently unstable from internal contradictions: if all morality is subjective, arbitrary, and relative, then why wouldn't those characteristics apply to the attendant claim that one should not impose one's morality on others? Accordingly, relativism undermines itself. Hence, sooner or later (at least for reflective adults, postmodernists notwithstanding)…the internal contradictions of meta-ethical relativism should eventually "disequilibrate" or perplex the thinker and prompt a "reequilibration" leading to movement beyond unbridled relativism to the achievement of post-skeptical rationalism. (Gibbs, 2014, pp. 87–88)

Finally, Kohlberg posited that such rationalism would give way to universal moral judgments in the highest levels of moral development, which are reached only by a minority of individuals. "Even in the adult years, only 13% fully or partially reached [the highest stages], and all of those had some graduate education" (Gibbs, 2014, p. 90). As Gibbs suggested, this is extremely problematic "because any theory-defining level, even in broadened form, misrepresents moral judgment maturity as the exclusive province of [those who are] philosophically or theoretically articulate" (p. 90).

Hoffman (2000) did not dictate a hierarchical or invariant sequence, beyond basic assumptions of human growth and development. However, nothing (to my understanding) about Hoffman's (2000) theory suggests the impossibility of a cognitively influenced tendency toward specific developmental results as Kohlberg posited. In fact, co-occurring cognitive development is assumed throughout the model. As Hoffman (2000) also explicitly rejected relativism, some ideal result of the developmental process is implicit (e.g., objectively correct moral judgments).[4] As Berman (1997) suggested, children selectively appropriate the material provided to them and their moral judgments are not entirely dependent on socialization. Thus, the theories are not a match, but neither are they incompatible. I see both deontological and consequentialist theories as important pieces in a puzzle that remains not entirely solved. In a complex physical and social world, it is certain that multiple influences, including and beyond those that we have explored, influence the development of our sense of right and wrong.

Finally, once individuals form moral principles (described in the following section), they begin to commit to them. Commitment represents internalization, self-image, and a sense of responsibility associated with moral principles

(Hoffman, 2000). Hoffman (2000) noted that during adolescence, abstraction and organization processes are significantly reconsidered when "one constructs one's own set of general, relatively abstract, though emotionally charged moral principles" (p. 260). However, once one has committed to and internalized moral principle, those principles are relatively stable.

Profound confrontations that test our commitment to moral principles are referred to as triggering events (Hoffman, 2000). Triggering events may produce emotional responses that cause one to reexamine their life choices and may produce new moral perspectives and a sense of responsibility (Hoffman, 2000). Such triggering events may threaten an individual's self-image (Hoffman, 2000). In the case of the Holocaust, some rescuers expressed that they could not have lived with themselves if they had not helped (Oliner & Oliner, 1988). However, while commitment to moral principle may motivate consistent behavior, it may also encourage individuals to interpret their own behaviors in a biased way in order to maintain a perception of moral consistency and reduce cognitive dissonance (Clayton & Opotow, 2003). Many non-rescuers interviewed by Oliner and Oliner (1988) were appalled by events that they witnessed during the Holocaust but felt that there was nothing that they could do. One possibility is that those non-rescuers maintained that belief as a defense against threats to their own self-image after inaction. I do want to be clear that I do not mean to condemn non-rescuers. Simply stated, it is easy to believe that one is the type of person who would have been a rescuer, but few are faced with a conflict of such profound significance wherein a definitive moment puts that self-conception to the test (Hoffman, 2000). Catalyzing action is discussed further at the end of Chapter 5.

In summary, children develop their sense of morality through socialization, during which they come to generalized moral conclusions through empathic arousal, associated feelings of guilt, and induction. This process is facilitated by the cognitive development process and furthermore children may come to a logical understanding of morality that is independent of socialization as a direct result of the development and function of their brain. They organize these moral concepts into a moral framework that becomes internalized and relatively stable. Beyond that point, the triggering experience may facilitate reassessment of one's moral principles.

Before continuing to discuss specific moral principles as outcomes of the moral development process, I briefly consider the influence of school and the influence of parenting on children's development. The implications of both digressions are discussed further in Part III, Pedagogy for Restoration. However, I feel that both influences may be significant to the development process and thus warrant consideration here.

Influence of parenting

The guardian(s) or parent(s) who raises a child was considered by Oliner and Oliner (1988) to have the greatest influence on the child's moral development (following Bowlby's attachment theory). Some go so far as to suggest that one's personality is largely determined and stable from a very young age and this position remains prominent today (e.g., Nave, Sherman, Funder, Hampson, & Goldberg, 2010). However this is not universally accepted, neither is it always interpreted deterministically, but it does warrant some consideration of the influence of parenting on moral development despite the focus of this book on education. Furthermore, much of the insight derived from Oliner and Oliner's (1988) consideration of parenting may also apply to teaching. Specific insights on parenting also serve as contextual understanding for educators in a schooling environment.

The use of power assertion by parents, guardians, or other authority figures is significant in a child's character development (Berman, 1997; Freire & Freire, 1992/1994; Hoffman, 2000; Oliner & Oliner, 1988). As explored in preceding sections, power assertive intervention by an authority in children's conflicts does not promote the use of induction, which is vital for moral development. Such intervention emphasizes obedience, an external locus of control, and does not facilitate learning of skills to cope with and resolve future conflicts. Oliner and Oliner (1988) suggested significant developmental consequences of excessive power assertion, which would cause children to become "constricted."

Constricted persons are described as those who

> are centered on themselves and their material needs. They tend to see relationships in terms of exchanges of material goods and benefits rather than in terms of connectedness or ethical responsibility. Their failure to act on behalf of others is an expression of their tendency to distance themselves from relationships that impose burdens on them. In the social and political arena they tend to define the world in terms of hierarchy of power and feel powerless to influence change. They tend to be passive and externalize blame for their circumstance. (Berman, 1997, pp. 88–89)

By contrast, extensive personalities are described as those who

> increasingly [incorporate] standards for personal integrity and care within their own value systems. While they may articulate such standards as cognitive principles, they experience them viscerally. They provide an organizing framework for their life activities and assignments of right and wrong. Even minor infractions distress them, and fundamental violations threaten them with a sense of chaos. (Oliner & Oliner, 1988, p. 250)

Less power assertive and more inductive parenting techniques were considered more conducive to extensive personalities (Oliner & Oliner, 1988). Extensive persons were posited to have what Berman (1997) called "leaky margins" where "ego boundaries sufficiently broadened so that other people were experienced as part of the self" (Oliner & Oliner, 1988, p. 183). Oliner and Oliner (1988) considered extensivity a quality conducive to altruistic behavior. Einolf (2010) followed Oliner and Oliner's (1988) work and attempted to measure extensivity and correlate it with prosocial behavior. Einolf (2010) found "quantitative evidence to support…the validity of the construct of extensivity, and the theory that extensivity motivates prosocial behavior." Notably, Einolf (2010) correlated extensivity with prosocial behavior less profoundly altruistic than the actions of the Holocaust rescuers such as donating time or money to charity.

I see constricted or extensive personalities as two extremes of a conceptual spectrum within which all people exist. I believe that Oliner and Oliner (1988) offer these archetypes as a way to recognize the influence of power and the importance of the use of induction for children. However, Oliner and Oliner (1988) did not find that those involved in rescue efforts were exclusively extensive (see the section "Conclusion: It Takes All Types" in Chapter 5).

I do not suggest that parents should be informed by academic theory in order to avoid raising constricted children. I posit that all parents, by social, moral, and evolutionary dictate, want the best for their children. In this regard, Freire and Freire (1992/1994) shared an important lesson on the application of moral development theory to the practice of parenting. Freire and Freire (1992/1994) had observed "the almost complete absence, not only of violent corporal punishment, but any punishment of children" (p. 13) in coastal fishing communities in Brazil and compared it with the emphasis on corporal punishment in inner-city Recife. Freire and Freire thought that the way of life in the fishing communities "had some connection with the taste for a liberty diametrically opposed to the use of violent punishment" (p. 13). Freire and Freire (1992/1994) considered this to be a "problem of the relationship between authority and freedom, which would necessarily involve the question of punishment and reward in education" (p. 22). Inspired by Piaget's theory on moral development, Paulo Freire eagerly gave a talk on the role of power assertive parenting in fighting oppression. In response, a member of the audience, whom Freire and Freire (1992/1994) describe as a peasant, said:

> It's one thing to come home, even tired, and find all the kids bathed, dressed up, clean, well fed—not hungry—and another to come home and find your kids dirty, hungry, crying, and making noise. And people have to get up at four in the morning the next day and start all over again—hurting, sad, hopeless. If people hit their kids, it's not because people don't love their kids. No, it's because life is so hard they don't have much choice. (p. 25)

Freire and Freire (1992/1994) call this "class knowledge" and described how this lack of understanding had offended the audience.

> It was the culmination of the learning process that I had undertaken long ago—that of the progressive educator: even when one must speak to the people, one must convert the "to" to a "with" the people. And this implies respect for the "knowledge of living experience" of which it is possible to go beyond it. (p. 26)

Freire and Freire (1992/1994) were reiterating the importance of popular education which Freire himself had first written about in *Pedagogy of the Oppressed* (1970). This concept is explored further in Part III.

The lesson that Freire (1994) shared speaks to the necessity of attention to social context. I don't suggest the man in the story was advocating for parents to hit their kids, but that "hurting, sad, hopeless" people see few good options. The man communicated how deeply insulting it is to insinuate that parents don't know or want what's best for their children and that someone unfamiliar with the context of their own lives could prescribe solutions. A similar sentiment, often expressed in regard to education, is that a student's parents don't care about their child's education. However, those who make such accusations rarely consider the context of the families they refer to. Again, I posit that all parents care about their children's education, but that many parents are faced with few good options when attending to their family's basic needs takes priority over school. Furthermore, schools may not be providing the education that parents need or want for their children. We learn from Freire and Freire (1992/1994) that theory doesn't simply and universally apply to the real world. It must be applied and adjusted in the context of a relationship with those with whom we work.

In summary, parenting is influential in the moral development of children. Increasing the use of inductive methods of discipline and reducing power assertion and especially punitive discipline almost certainly, but not exclusively, fosters extensive personalities. In turn, these children may be more likely to participate in democracy, to help others, and generally to be engaged citizens (Berman, 1997). This is not, however, the only path to such behavior. Finally, as learned from and by Freire and Freire (1992/1994) it is not enough simply to state that parenting should be done "better" and to assume that parents don't know or want what is best for their children. An extensive society is conducive to extensive parenting and this is where the discussion of parenting ends and the discussion of a restorative pedagogy begins.

Influence of school

Hoffman (2000) discussed the changing emphasis of moral principles in schooling that takes place at about the fifth grade. At that time, the earlier emphasis from parents and early primary schooling shifts from self-improvement to social comparison. Before the shift, caring and achievement apply in different contexts. After the shift, caring and achievement often apply in the same context and conflict with each other because of the competitive nature of school and society in the United States.

> The principles of "equality" and "need," associated with caring early in life, are largely left behind; the shift to social comparison adds competitiveness to peer relationships and prepares children cognitively and emotionally for the merit based justice orientation that will dominate their adult lives. (Hoffman, 2000, p. 258)

This led Hoffman (2000) to question the impact this shift has on children:

> That children are socialized initially to share but later taught to value merit suggests a discontinuity in justice socialization: children must begin to unlearn "equality" in favor of "merit." We can only guess the long-term effects of discontinuity between "equality," intense motivation to compete, and achievement guilt, until research provides more definitive answers." (p. 258)

I appreciate that Hoffman (2000) identified this discontinuity and wish it had been explored further. As Gibbs (2014) argued, descriptive accounts may downplay the severity of issues and become value specific in their attempt at neutrality. The psychological, philosophical, and moral nature of moral research dictates that empiricism must be coupled with wisdom and furthermore must be value specific and always focused on improving life. This means that moral researchers must focus on prescription as well as description.

While more knowledge, especially in moral conflicts, might be desirable, our knowledge is already practically infinite (Haidt, 2006). We collectively know more than anyone can know individually and we certainly know enough to be doing things better. If the drive for empirical knowledge is simply a vocation and not the pursuit of wisdom that would improve lives and furthermore prevent action through the endless call for more research, then empiricism maintains the status quo and is a tool of oppression. It would be true that, as Horkheimer and Adorno (1944/2002) wrote, "on their way toward modern science human beings have discarded meaning" (p. 3). Indeed, such empiricism would be one of the master's tools. However, if such empiricism contributes to a greater understanding and to wisdom and action, then it can be a tool of liberation. This is not to suggest that no one in or beyond the field is doing anything, but that we all need to do more. Neither do I intend to condemn Hoffman (2000). If anything, my critique is aimed at modernism and academia.

I find this to be the conflict that I have struggled with throughout my life as a student and especially in the course of writing this book. The analysis paralysis that has plagued my life as a graduate student is a source of constant frustration and not actively working for what my research dictates is immensely frustrating. I imagine that the frustration Berman (1997) expresses to such issues may be similarly motivated:

> With current educational practices focused primarily on individual competence within a competitive society, treating social responsibility in most school districts is a rhetorical slogan rather than a practical reality. Current methods of instruction and current classroom and school structures promote obedience and alienation rather than thoughtfulness, engagement, and commitment. We have lobotomized children, separating their social consciousness from their individual identity. (p. 185)

We are actively harming our children and investigations into how we are harming them are important, but it is crucial that we actively work to reduce and stop the source of harm. If we do not take action, we might very well continue to identify the conditions that will lead us into oblivion as we march into the abyss. This concern is revisited in the conclusion of this work.

Moral Principles

Thus far, human morality (including empathy) has been considered a fundamental component of humanity. Human morality encourages us to cooperate and to resolve conflicts between our own self-interest and the interest of others without harming one another. Obviously, conflict resulting in significant harm continues to occur in human civilization (see the section "Limitations of Empathy" in Chapter 3 and the section "Social Death" Chapter 5). The fact cannot be ignored that some act with malice or without regard to the interests of others (see the section "Outlying Behavior" in Chapter 5). However, aside from such outliers and deviants, most of us develop a set of principles which act in consideration of (if not accordance with) the interests of others.

As already stated, Hoffman (2000) generally conceptualized moral principles as psychological outcomes of moral development through affective socialization (consequentialism), while others see them as purely philosophic and absolute concepts (deontology). However, the principles themselves are significantly cognitive. Once empathy and moral principle "bond," the cognitive task of identifying or recognizing violations of justice can trigger affective or empathic responses. The bonding of affect and cognition that Hoffman (2000) described is referred to as "hot cognition." Moral principles are stored in long-term memory and require more cognitive processing, more time for retrieval, than affective empathy (e.g., mimicry, conditioning, and direct association). Processing empathy, a nearly instantaneous process, can be a powerful retrieval cue for moral principles (Hoffman, 2000).

Empathy and upholding moral principles have the capacity to motivate prosocial action and in tandem often motivate people to a greater degree than either alone. However, it is also possible for empathy and upholding moral principles to conflict, causing diminished motivation (Decety & Cowell, 2014). In some cases, empathy for an individual may overcome considerations of fairness for a collective. This is certainly evident in the human tendency toward nepotism (see the section "Moral Exclusion" in Chapter 5). Thus, the following exploration of moral principle should be considered in relation to human empathy.

In this section, the moral concepts of rights, caring, and justice are explored. Three principles of justice are presented: merit, equality, and need. Next, social and ecological implications of each concept are briefly considered. I take the position that none of the concepts are naturally more or less prosocial, but that they can be interpreted in myriad ways. The following moral principles motivate both obligatory and discretionary judgments. If any objective morality does exist, it seems that there must be multiple ways of coming to such universal judgments.

Rights

Rights are the fundamental set of rules regarding how we treat one another. Rights may be the most fundamental moral principle underlying all moral judgments. Most individuals, in the context of their own culture, have an intuitive sense of rights that they hold to be universal. The central consideration in this discussion of rights is whether universal rights exist or whether they are simply social conventions. The implications of universal rights versus those of rights dictated by consensus are then considered. Finally, I identify specific rights that are either universal or nearly universally agreed upon social conventions.

Rights must protect and limit self-interest and collective interest. Universal rights (e.g., human rights) apply equally to everyone, while conventional rights are specific to the context of a particular group (e.g., civil rights). The ideal role of conventional rights is both to protect universal rights and to establish specific context-dependent rules for society, but the distinction is not always clear. Conventional rights may be socially constructed in consideration of philosophic ideals of universal and inalienable rights. For example, the right to bear arms is not guaranteed to all people in all societies. It does, however, relate to what many consider to be a universal right to liberty. However, most humans consent to limitations of their personal liberty and the personal liberty of others in order to promote and optimally protect the liberty and safety of all. It seems that an unqualified human right to liberty is neither universal nor conventional.[5]

Many philosophic approaches to rights attempt to identify the nature and universality of rights. A foundationalist approach to rights seeks to find a basis for universal human rights beyond social construction and consensus (Schapiro, 2005). One such philosophic approach suggests that the laws of nature dictate certain rights (natural law and natural rights). In this philosophic tradition, natural law is considered the law of God. Others have sought ecologically or biologically dictated foundations for rights (Fruehwald, 2010). Many reject a foundationalist approach entirely and prefer practical considerations of rights in order to establish a baseline for human by consensus

(Freeman, 1994). Some even suggest that the only reasonable basis for rights is to construct rights in reaction to historical violations. Finally, others see rights as entirely subjective social conventions with no possibility of universality.

Although a deep exploration of the foundations of rights cannot be accomplished here, I explore such a possibility very tentatively as an illustration of foundationalism. I make a basic assumption throughout this book that all life is worthy of life. A cursory interpretation of such an assumption would be that every living thing has the right to continue living. Predation and natural causes present obvious exceptions to that right. However, natural selection dictates no natural right for a species to exist. In fact, species who do not adapt to environmental changes or who are outcompeted become extinct. This becomes a treacherous position as a foundation for human rights because it may be that there is no foundational right for any species to live.

Obviously, the above was a hypothetical example; however, it illustrates the controversy over a foundational approach to rights deferring to a higher power (natural selection, cognition, God, etc.). The foundational approach may call into question or even deny practical rights that are clearly prosocial. However, the opposite extreme at which no foundation can be accepted may suggest that rights are entirely normative and contingent. Lack of a foundation for international human rights efforts may call the validity of the judgments of human rights organizations into question when identifying human rights violations. Practical investigation of the function of rights in consideration of one's own moral position offers more utility. If a right functions to protect human welfare regardless of context, it could be considered a universal right.

To return to the central question introduced at the beginning of this section: Are rights universal? There can be no simple conclusion.

> A conception of human rights should be flexible enough to allow space for the human creativity it seeks to defend and to address the changing conditions of the world that may threaten its values.... One must make a nonrational decision either to accept or reject solidarity with humanity. (Freeman, 1994)

It is not the identification of universal rights, but the commitment to protecting and respecting the welfare of others that must take precedence. The only entirely and unquestionably universal right is the right to be morally considered. Thus, I favor a universal ethic of care to universal considerations of rights. However, this is no excuse not to construct and consent upon practical rights. As in other pragmatic considerations of human potential throughout this work, I suggest that rights should be iteratively consented upon, evaluated, and reassessed at the individual level, among communities and nations, and globally in accordance with changing conditions on Earth.

The United Nations' *Universal Declaration of Human Rights* may be the most prominent effort toward reaching universal rights by consensus (United Nations, General Assembly, 1948).

> Jacques Maritain, a leading French philosopher and one of the participants in the negotiations that resulted in the Universal Declaration of Human Rights, illustrate[d] [the] nonfoundational approach. In response to expressions of surprise at mustering widespread agreement to the Universal Declaration, Maritain stated…"yes, we agree about the rights but on condition no one asks us why." (Schapiro, 2005)

Maritain was acknowledging the fact that the authors and signatories were motivated by differing theoretical considerations but found consensus in the function of rights. The *Declaration* makes no appeals to God or the law of nature. Indeed, such explicit foundations would be grounds for much debate and empirical criticism. Instead, it is an entirely functional and context-independent document proposing freedom and dignity for all humans—subject to broad interpretation. Some of the rights dictated by the *Declaration* are the right to life, dignity, security of person, and the denouncement of slavery, torture, arbitrary arrest, detention, or exile (United Nations, General Assembly, 1948).

Finally, I want to consider the possibility and implications of extending the concept of rights beyond humans. Based on my previous analysis I suggest that rights are a human construct. I don't dismiss the possibility of natural rights that would apply to non-human animals, or the possibility of such formalized moral concepts among other animals. However, regardless of foundation, human conceptions of universal rights should apply to our consideration of non-human animals and Earth. Most human societies do establish conventional rights (e.g., laws) for the treatment of animals. We similarly have laws and agencies dedicated to protecting natural environments. A universal declaration of animal rights or planetary rights may not exist as such, but many international agreements do protect "rights" of animals and the environment. For example, the International Whaling Convention universally banned commercial whaling. I propose that human concepts of rights exclusively apply to human behavior. For example, sharks do not observe the internationally consented upon right of whales not to be hunted. Thus, animal rights would be unnecessary in the absence of humans not because non-humans lack intrinsic value, but because animal rights focus on protecting non-humans from humans.

In summary, I suggest that rights are nearly universal. In the complexity of human social interaction and environment, rights can be neither static nor context-independent. However, promoting and protecting human welfare (and the welfare of non-human animals) by establishing a set of near-universally consented upon rights serves a great function in society. For example, someone may observe another's right to security of person despite being very angry at that person and potentially even feeling that hurting that person could be justified based on considerations of the other's merit. Thus, although rights are impersonal, such general agreement on rights moderates the potential subjectivity of individual moral judgments (see the section "Cultural Normative Limitations" in Chapter 5). By establishing rights, certain harmful behavior on the part of individuals or governments is universally discouraged. In the resolution of conflict, folks likely consider multiple moral principles in their judgments.[6]

Care

Care extends beyond rights that may be considered morally obligatory into the realm of "goods" (discretionary judgments). Rights encourage consideration and protection of one's basic needs while care encourages both broad consideration of well-being and attention to individuals' needs. Hoffman (2000) defined caring as a somewhat utilitarian (consequentialist) principle that aims to accomplish the greatest good for the greatest number. However, Oliner and Oliner (1988) defined an ethic of care as "a dominating sense of obligation to help all people out of a spirit of generosity and concern for their welfare" (p. 217). I see care as a potentially universally prosocial moral principle dictating that everyone deserves to have their basic physiological and psychological needs met. Thus, care is a mature concept of reciprocity and neglect and abandonment are moral violations of care because everyone has a duty to help anyone in need. Oliner and Oliner (1988) indicated that 76% of the interviewed rescuers were motivated by an ethic of care.

> The ethical values of care and inclusiveness that distinguished rescuers were not merely abstract or philosophical preferences. Rather, they reflected a key dimension of rescuers' personalities—the way they characteristically related to others and their sense of commitment to them. (Oliner & Oliner, 1988, p. 169)

The following are some of the sentiments expressed by those rescuers:

> I think if somebody is in a bad situation, you have to help them. (p. 170)
>
> We had to give our help to those people in order to save them. Not because they were Jewish, but because every persecuted human needs some help, just as my father found help when the Turks killed the Albanians.[7] (p. 169)

One rescuer who risked his life to help a Nazi soldier said that "the moment the man was badly wounded, he was not an enemy anymore but simply a human being in need" (p. 288). This rescuer was certainly motivated by care and empathy. An observer may be empathically motivated to help someone in need and also feel obligated to help based on a principle of caring (Hoffman, 2000). Affectively motivated induction may have also been fundamental to the formation of one's principle of care. However, a principle of care alone, in the absence of significant empathy, has been found to be a sufficient prosocial motivator (Oliner & Oliner, 1988). Hoffman (2000) concluded that "empathy and caring principles are thus independent, mutually supportive, hence congruent dispositions to help others" (p. 255).

Finally, an ethic of care easily extends to non-human animals and the planet. Animal shelters, sanctuaries, and rehabilitation centers represent an ethic of care for animals. Such efforts tend to focus on animals valued by humans based on our perceived similarity, cuteness, or in the interest of preserving biological diversity. However, this is not exclusively so, and some animal protection efforts may exclusively represent an ethic of care. It becomes more difficult to gauge the role of self-interest in humans on ecological protection efforts; however, many do see intrinsic value in nature.

Justice

Considerations of justice are often made in consideration of rights and care that dictate that a basic standard of behavior and treatment of others applies broadly. However, the primary concern of justice considerations is to determine what is fair. Justice considerations potentially reflect more immature reciprocity; however, mature reciprocity is not incongruent with any justice principles. I focus, as Hoffman (2000) did, on distributive justice and the principles of merit, equality, and need. Hoffman (2000) described distributive justice as the correlation between reward and deservingness in the distribution of resources. However, Clayton and Opotow (1993) conceptualized that distribution of resources is not strictly material but includes "such socially-valued goods and resources as love and caring, services, goods, money, information, and status" (p. 300). Thus, other conceptions of justice such as punitive and procedural justice are readily considered within a broad conceptualization of distributive justice.

Punitive (or retributive) justice is readily examined as the distribution of punishment wherein relations of power are particularly influential. Punitive justice is largely based on an immature form of reciprocity wherein the severity of punishment is reflective of the egregiousness of the offence. However, the Māori concept of restorative justice provides an alternate to punitive justice. Restorative justice is focused on making amends for transgressions and restoring justice rather than punishing injustice. In the restorative justice process, the victim and offender meet with their families, community, and facilitators to have a discussion about the incident. The offender is provided the opportunity to make restitutions, restore damaged relationships, and potentially be accepted back into the community. The process of restitution is not predetermined by the facilitators, but the offenders may voluntarily offer apologies, gifts, or money.

In the documentary *Restoring Hope: An Indigenous Response to Justice*, Mike Hinton, a restorative justice facilitator in South Auckland, said:

> Restorative justice isn't about forgiveness, alright? That's a byproduct of what happens at the conference. It's about communication. And I'm probably gonna' upset a few people by saying this: an eye for an eye probably doesn't work that well. And maybe it's a practice that we should not look at. Maybe something like communicating, understanding, respecting, and acknowledging differences; taking responsibility for hurt and harm, and making amends would work a little bit better, today. We can grow as people. Our process helps people grow. It helps them heal and grow. (Carnachan & Parnell, 2013)

By denouncing "an eye for an eye" ("immature" reciprocity), Hinton illuminated restorative justice as a form of "mature" reciprocity. In New Zealand, the restorative justice process occurs in conjunction with the penal system and outcomes of the restorative justice process are considered by courts, but some offenders also face jail time. However, the intention of restorative justice is to restore balance, heal, and move forward rather than to punish the offender.

Similarly, distributive justice relates to the fairness of social interactions and decision making and can be considered distributive because distributive justice begins with the distribution of membership into a community (Opotow, 1990). One's membership in the community (in the moral sense) influences whether or not they are considered to have the rights of community members including moral consideration (see the section "Moral Exclusion" in Chapter 5). Thus, I proceed with considerations of justice based on principles of distributive justice.

Oliner and Oliner (1988) stated that approximately 15% of the rescuers they interviewed were motivated primarily by a broad sense of justice congruent with Kahn's (1999) definition of obligatory morality and significantly cognitive. The following are some of the sentiments expressed by those rescuers:

> I found it incomprehensible and inadmissible that for religious reasons or as a result of religious choice, Jews would be persecuted. It's like saving somebody who is drowning. You don't ask what God they pray to. You just go save them. (p. 166)

> Jewish people had as much right to live as I did. (p. 167)

> All men [*sic*] are equal and are free and equal by right. Consequently I am against all dictatorial systems. (p. 167)

The above quotes appeal to "fairness" based on appeals to rights and caring. Other justice considerations appeal more clearly to the specific principles of justice; however, generally judgments of fairness are made in consideration of multiple principles.

A merit-based principle of justice reflects consideration of one's deservingness in determination of fairness. The basic sentiment is that "you reap what you sow." Hoffman (2000) finds merit to be the dominating moral principle in modern Western culture. In the distribution of goods, merit may dictate a capitalist system in which some deserve more material wealth than others based on their contribution to the economy. In a more procedural sense, the merit of one's behavior may determine what is considered fair treatment. In the most severe cases, judgments of merit may be used to strip individuals of their rights. Merit-based considerations may be conducive to victim blaming. For example, if someone is suffering, one may judge that the person either has not worked hard enough or has done something to deserve the harsh circumstances. Although I have raised potential pitfalls of merit-based considerations, justice considerations based on merit are not intrinsically non-prosocial, especially when considered along with principles of rights and care. It is prosocial to recognize positive contributions of others. Humans may not be so intrinsically motivated that we would all strive for "goodness" in the absence of any external consideration of merit. I argue only that merit must be moderated by other moral considerations.

An equality-based principle of justice dictates that everyone deserves the same rights, resources, and rewards. In the distribution of goods, equality may dictate a communist system in which everyone deserves equal material wealth. In a more procedural sense, judgments of equality would likely defer to appeals of equal rights and dictate that everyone should be judged on the same basis with less consideration of personal experience and needs. It is important to understand the difference between equality and equity. Equality suggests that everyone deserves the same thing, while equity is a more general concept of fairness in consideration of the individual needs. Equality, like rights, may sometimes encourage impersonal judgments. In consideration of equity, equality it is not necessarily universally prosocial to treat everyone equally.

A need-based principle of justice dictates that people deserve to have their basic needs met regardless of their personal merit or contribution. This is similar to the moral voice of care. An ethic of care would likely encourage considerations of justice based on need, but care is a broader concept of morality and need is specifically related to justice. In the distribution of goods, need may dictate a socialist system wherein the emphasis is on meeting individuals' basic needs (e.g., universal health care!). The basic sentiment is encapsulated by the slogan popularized by Karl Marx: "From each according to his [*sic*] ability, to each according to his [*sic*] need."

Merit, equality, and need represent a spectrum of the emphasis of fairness considerations from deservingness (merit) to entitlement (need). Prototypical examples of an emphasis on a particular principle exist, but I have argued that moral judgments usually reflect more complex considerations involving multiple principles. Furthermore, a particular situation may encourage emphasis on different moral principles depending on the nature of the interaction and the relationship between the parties (Clayton & Opotow, 2003). For example, in close personal relationships, a strict demand for reciprocity may communicate a lack of value of the relationship (Clayton & Opotow, 2003). Merit may be more congruent with a strict demand for reciprocity (immature reciprocity) and need with mature reciprocity; however, this is not exclusively so.

As already noted, moral principles and empathy have the potential to enhance or diminish each other (Decety & Cowell, 2014). As Hoffman (2000) explored, the prominence of a particular principle and its congruence with empathy depend on the situation. Consider two examples: (1) A classmate asks you for answers to an assignment on which you worked very hard. (2) You witness a drunk driver cause a collision. In the first case, some might help a classmate without question out of a sense of care. Others might consider their classmate's merit; it would make a difference whether that classmate had made an effort or had spent the previous night drinking beer and playing video games while you spent the entire night on the assignment. Even if one lacked merit, you may be persuaded by need if the classmate laments that they risk failing the class. Finally, the classmate may make an emotional appeal, causing you to help despite it being unfair. Helping someone with homework is potentially good (discretionary judgment) but not obligatory. In a situation of greater gravity, such as the second example in which a drunk driver may be in significant danger, that person's clear lack of merit may be less influential in one's decision to help. One may simultaneously believe the driver deserves punishment and that they deserve to be helped regardless of their merit depending on the severity of the accident.

Justice considerations shed more light on the complexity of judgments regarding ecological degradation. One must judge whether animals and the planet deserve consideration (explored in the section "Moral Exclusion" in Chapter 5) and balance human interests with ecological interests. These considerations become much more subjective than universal rights or care for non-humans. For example, considerations of merit may determine that some species or habitats are more worthy of consideration and protection than others. Similarly, one may consider humans superior because they believe human interest always trumps ecological interest.

· 5 ·

SOCIAL INFLUENCE AND INDIVIDUAL DIFFERENCE

> So convenient a thing is it to be a reasonable creature, since it enables one to find or make a reason for every thing one has a mind to do. (Benjamin Franklin as cited in Haidt, 2006, p. 66)

Just as empathy as a prosocial motive has limitations, so does moral principle. I choose to believe that very few humans deliberately engage in antisocial behavior or otherwise deliberately cause harm to one another without justification (see the section "Outlying Behavior"). Instead, I believe that the process of moral development, in the context of social norms, influences the primacy of conflicting moral principles, conception of moral encounters, and moral judgments and actions.

Turiel (2005) discussed the events surrounding the integration of the University of Mississippi in 1962 and suggested that those either supporting or opposing the integration of the university were acting on commitment to principles. It would be easy to suggest that Governor Barnett's attempt to block integration of the university and causing President Kennedy to bring in the National Guard (more than 30,000 troops) were symptoms of some gross iniquity prevalent at that time in the U.S. South. However, Turiel (2005) considered Governor Barnett's defense of segregation to represent moral "conviction, commitment, and courage" (p. 17). Similarly, James Meredith, an African American man, struggled for equality and integration displaying the same (I believe greater) "conviction, commitment, and courage." Meredith took his case to the U.S. Supreme Court, whose justices ordered his admission into the University of Mississippi (Turiel, 2005). Although Turiel does not does not explicitly refer to either Meredith's or Barnett's actions as moral, I want to consider the possibility that both were acting morally.[1]

It is possible that Barnett was acting morally. But first I want to consider the possibility that he wasn't. Barnett expressed pride as a Mississippi segregationist and defended his position, but it is impossible to understand his motivations based on historical accounts. In many cases, the appearance of moral conviction is more important to an individual than is actual moral conviction (Haidt, 2006) and that must be doubly true for politicians. Furthermore, humans have a tendency to fabricate post hoc explanations and justifications for our behavior, often without actually understanding our motivations (a process called confabulation). We are not necessarily even aware that our justification or explanation was not the true motivation of our behavior (Haidt, 2006). Additionally, people have a self-serving bias wherein we often see our own actions through "rose-colored glasses," justifying actions that we would almost certainly condemn if carried out by anyone but ourselves. There is no reason to assume Barnett was acting morally.

However, Barnett may have been acting morally. How could it be possible that Meredith, an African American man fighting for civil rights, and Barnett, a white man fighting to defend the discriminatory practice of segregation, were both acting morally? Regardless of whether Governor Barnett acted with conviction, commitment, and courage, it was not the same as Meredith's. Although the most deplorable tyrants and heroic liberators of history may be similarly motivated by a sense of partisanship and their actions judged through a particular socio-historical lens, the tyrant co-opts justice while the liberator loves and serves justice. If anything could ever be considered universally wrong, it must be oppression. However, much of this chapter explores how and why moral behavior may fall short. My presumption is that the majority of people do act morally, but that the way we perceive and respond to moral conflicts is very subjective. I neither accept morality as universal nor do I resign to the permissiveness associated with moral subjectivity. For now, I simply suggest that moral behavior can be harmful.

Moral Exclusion

In the case of Governor Barnett and James Meredith, it is likely that Barnett did not believe that, as an African American man, Meredith deserved the same rights as whites. Such exclusion may be justified by a principle of merit. Thus, the concept of moral exclusion sheds some light on the case.

> Moral exclusion occurs when individuals or groups are perceived as outside the boundary in which moral values, rules, and considerations of fairness apply. Those who are morally excluded are perceived as nonentities, expendable, or undeserving; consequently, harming them appears acceptable, appropriate, or just. Moral exclusion…links a wide range of social issues, such as abortion, species conservation, nuclear weapons, and immigration policies, because our position on these issues depends on whom we include or exclude from our moral boundaries. (Opotow, 1990)

Opotow (1990) found three "attitudes that comprised moral inclusion: (1) believing that considerations of fairness apply to another, (2) willingness to allocate a share of community resources to another, and (3) willingness to make sacrifices to foster another's wellbeing." These attitudes are all compatible with Hoffman's (2000) theory on moral development driven by guilt. However, Opotow (1990) offers significant insight into how human boundaries of moral exclusion and inclusion expand and contract. These borders change with the status quo and with conflict, and are not absolute, but better thought of as degrees of inclusion or exclusion.

One of the psychological origins of moral exclusion is conflict. Danger of any type, conflict, and stress "reinforce[s] group boundaries and change[s] information processing strategies and choice of justice rules" (Opotow, 1990). With conflict, group cohesion increases and concern for justice to those outside of the group decreases. Outsiders may be increasingly endangered and subject to extreme forms of oppression such as exploitation, slavery, and extermination, but the less overt forms of dominance should not be overlooked. In situations of conflict, resolution looks very different between folks who consider one another to be within their own scope of moral inclusion. In these cases, conflict is "regulated competition of equals, conducted according to rules of fair play, such as a duel or bidding war; with those outside, conflict is unregulated, no-holds-barred power struggle among unequals, such as guerilla warfare" (Opotow, 1990). Conflict also tends to constrict one's borders for moral inclusion, while diminished conflict offers opportunity for the borders to expand (Opotow, 1990). This is consistent with the in-group empathic bias and its evolutionary hypothesis that conflict over competition for resources in human evolution produced preferential selection of human traits favoring in-group bias. However, this suggests that the bias is not hardwired in our DNA, or even explicitly social, but a result of immediate and present conflict. It seems that through efforts to reduce conflict, such as reparation, compensation, or simply helping folks meet their basic needs, we would all create space for our moral borders to extend.

The other psychological origin of moral exclusion that Opotow (1990) discussed is unconnectedness. This stems from the innate human tendency to differentiate and categorize objects. When this is applied to social categorization, it may rationalize injustice by affirming a sense of separateness between a witness and someone experiencing injustice. Oliner and Oliner (1988) discussed how Nazis used different tactics in each country in accordance with the prewar status of Jewish people in those countries. For example, in the Netherlands, Nazi efforts to morally exclude Jewish people preceded efforts to enact genocide.

> The Nazis did not want to shock Dutch sensibilities and provoke widespread resistance. Instead, they gradually disenfranchised, impoverished, and isolated Jews in a period when German domination over Dutch gentiles was still relatively tolerable, thereby dissociating the latter as much as possible from Jewish suffering. (p. 133)

Opotow (1990) explained that in opposition to this sense of unconnectedness, perceiving someone as connected to oneself hinders moral exclusion. This may include "belonging to the same community, perceiving another as a worthwhile being, or discerning any thread of connectedness" (Opotow, 1990). This connection leads to increased empathy and ultimately helping behavior. This sense of connectedness is highly sensitive to conflict and increased severity of conflict corresponds with increased feeling of unconnectedness and reduces helping behavior (Opotow, 1990). Oliner and Oliner (1988) discussed the effects of social categorization independently of its potential for producing feelings of unconnectedness, stating that "both positive and negative stereotypes lend to thinking of Jews as a group rather than as individuals" and that "rescuers saw Jews more frequently as individuals" (p. 151). In Opotow's (1990) discussion of the intersection of psychological and sociological forces at work in moral exclusion, they conclude that "the bidirectional influence between individuals and society in perpetuating moral exclusion suggests possible ways to interrupt the cycle of harm" (Opotow, 1990).

Opotow (1990) reviewed various conceptions of the boundaries of moral inclusion, suggesting that one of the more prevalent conceptions is that the boundaries should be drawn around membership status of the human species. A prevalent outcome of this conception is debate over who can be considered human. Categorizing humans by race, class, and gender has been a common tool in the enactment of oppression to deny the humanness of a person, people, or peoples in order to deny their inclusion in moral considerations. Instead, Singer (1975) believed that the boundaries should be constructed around cognitive awareness and that nonhuman sentient beings deserve inclusion in moral consideration. Opotow (1990) also discussed the belief that all life should be treated with reverence, and thus considered to be within the moral boundaries as, for example, Jains believe. This begins to illustrate that societal, cultural, and religious norms heavily influence not only the construction of moral principle, but also its application. As discussed in the construction of moral principles, there is certainly an individual, non-deterministic element influencing how we define our own moral boundaries.

Moral exclusion and ecological issues

Many controversial issues can be considered to result from differing moral boundaries. The central issue in the abortion debate is also readily reduced to a consideration of moral inclusion. The point at which a fetus is considered worthy of moral inclusion and the impact that has on the degree of moral inclusion of the pregnant woman encapsulate all of the arguments in this debate from both sides (Opotow, 1990). Opotow also gave an example regarding whaling and suggested that many U.S. Americans are outraged by Japanese whaling practices because the moral boundaries of U.S Americans include whales to a greater degree than the Japanese. However, whaling has been a large industry in United States in recent history. Moral exclusion and differing moral boundaries provide insight into conflicts in other aspects of animal rights activism, as well, where animal rights activists are commonly criticized for caring more about non-human animals than about humans. As Plous (2003) argues, the process of excluding non-human animals from moral consideration psychologically parallels the process of excluding other humans.

The concept of moral exclusion and moral boundaries sheds light on the relationship between environmental and social degradation as well. Let's return to the study (discussed in Chapter 1) in which Kahn (1999) interviewed Black children in Houston, Texas, about their perceptions of nature. In response to a question about what the children thought about when they thought of nature, 7% mentioned drugs or violence and 17% reported that it was an environmental issue that they discussed at home (Kahn, 1999). I was struck by one child's responses to the interviewer's attempt to understand how they connected with non-human animals. The child said that they had a cat and the interviewer asked whether the cat was important to them. The child responded, "No. I have other things that's important to me. If I eat or not. Or if anybody in my family is gonna die, because I don't want nobody in my family to die" (Kahn, 1999, p. 109). The same child expressed similarly demoralizing concerns in response to the interviewer's other questions relating to the child's connection with and concern for nature. When the interviewer asked the child about parks, the response illustrated concern for dangers of broken glass and gun violence. It seems evident that such immediate concerns would preclude ecological ones.

The majority of children in the Houston study did provide moral justification for judgments relating to the environment. Most of the justifications were anthropocentric and the children expressed very few biocentric justifications. Kahn (1999) stresses that "anthropocentric reasoning should not be discounted" (p. 113) in either the larger context of environmentalism or in the context of the children in Houston. In fact, Kahn (1999) suggests that "the Houston children articulated forms of anthropocentric reasoning that many would argue are not only philosophically valid, but practically speaking—in terms of effecting societal change—an imperative" (p. 113). With the Houston study in mind, I suggest that anthropocentric and biocentric perspective taking and motives are equally important in the struggle for restoration and position the needs of the planet and the needs of humanity in congruence.

An alternate conception of moral justifications involves microjustice (focused on the individual) and macrojustice (focused on the group) (Clayton & Opotow, 2003). In environmental conflicts, macrolevel judgments favor the needs of the collective (Clayton & Opotow, 2003). Depending on the specific situation and who is morally included, neither macro- nor microlevel judgments are strictly anthropocentric or biocentric. Clayton and Opotow (2003) found that macrolevel considerations were more common for pro-environmental arguments and microlevel considerations, such as the rights of landowners, were more common in anti-environmental arguments. However, neither micro- nor macrolevel justifications deny consideration of self-interest because collective welfare is instrumental to individual welfare.

Although humans have the capacity for altruism, we must also consider our own self-interest to ensure that our basic needs are met.

This calls into question the nature of biocentric reasoning and has led some environmental philosophers to speculate that all human reasoning is anthropocentric (Kahn, 1999). In the attempt to understand how people are morally motivated, the micro/macro conception provides more insight. However, I have tried to embrace both/and thinking and micro/macro considerations easily incorporate biocentric/anthropocentric considerations. This does not diminish the value of a biocentric perspective (in distinction from strictly biocentric reasoning), which has been shown to be a powerful motivator for pro-environmental moral judgments and empathy (Berenguer, 2007, 2010). Nor, as Kahn (1999) points out, should anthropocentric perspectives be undervalued in their power as a prosocial and pro-environmental motivator.

Building on Kahn's (1999) earlier work, Severson and Kahn (2010) tested whether children's anthropocentric perspective or biocentric perspective taking was more dependent on the situation that they were considering than their values. Severson and Kahn (2010) posed a hypothetical situation wherein there was no anthropocentric stake in nature. Aliens had come to an Earth uninhabited by humans and caused harm to nature (including pets, wild animals, orchards, and forests). The children were asked whether it would be wrong for the aliens to do such harm and the majority answered that it was wrong to do harm to nature even without the presence of humans. This suggested that

> if children are asked questions about environmental harms when humans are involved, then biocentric reasoning appears seldom, and it appears late in development. But if children are asked questions about environmental harms without the pull of human consideration,...then biocentric reasoning appears often and early. (Severson & Kahn, 2010, p. 255)

With these perspectives on biocentric and anthropocentric reasoning and the micro/macro level judgments in mind, I will examine an example from Kahn's (1999) Houston study. When a child was asked whether it is all right to throw trash in the bayou, they answered:

> [It's not all right] because some people that don't have homes, they go and drink out of the rivers and stuff and they could die because they get all that dirt and stuff inside their bodies. (Kahn, 1999, p. 101)

This appears to be obviously anthropocentric but in consideration of the micro/macro model of justice, it is not so simple. This child may be demonstrating moral concern for the bayou at the macrolevel where the value of the bayou is for water as a collective resource. While the child's explanation centers on humans, specifically people experiencing homelessness, that doesn't necessarily morally exclude others (other people and the natural environment) from the collective. I believe that this illustrates only that the child was immediately concerned about homelessness and does not necessarily illustrate an anthropocentric orientation. As evidenced by Kahn and Severson (2010), their interviewee may have reflected a more significantly biocentric orientation if asked in a different way. Especially because the question relates to anthropogenic harm, it may warrant a more anthropocentric judgment.

A final consideration related to moral exclusion is denial of self-involvement (Opotow & Weiss, 2000). Individuals may consider their own actions unrelated to the well-being of others, not because they do not consider them worthy of moral consideration, but because they deny that their actions affect others. Through denial of self-involvement, some may believe that they do not pollute so they are not involved in the ecological crisis (Opotow & Weiss, 2000). "Another route to denial of self-involvement recognizes a problem as real but focuses on parts of the problem that exonerate oneself and therefore denies one's contribution to the problem" (Opotow & Weiss, 2000, p. 486). One such example might be someone's proud ecological commitment to drive a hybrid car 30,000 miles per year. Self-exclusion also applies to racism. Conceptions of self-exclusion may allow folks to believe that racism is a terrible, awful thing that, thankfully, they are not involved in, or that they have successfully put an end to, for example, a university's efforts to increase racial diversity while ignoring institutional racism.

Subjugation of animals

I haven't focused too intently on the subjugation of animals because I don't want to dismiss or trivialize the subjugation of humans. In consideration of the child in Houston who said she didn't care about her cat because she cared whether she would eat or not (Kahn, 1999), I think emphasis on animal rights and ecological rights are, for good reason, poorly received in the face of such severe violations of human rights. However, there are many parallels between the oppression of humans and the oppression of other animals. In this section,

I explore the relationship between the mistreatment of non-human animals and the mistreatment of humans.

Even if one does not consider other species worthy of moral consideration, humane treatment of non-human animals likely parallels our humane treatment of one another. Although Kant did not consider animals to have any intrinsic value, Kant recognized that by mistreating animals, humans degrade their own humanity.

> If a man shoots his dog because the animal is no longer capable of service, he does not fail in his duty to the dog, for the dog cannot judge, but his act is inhuman and damages in himself that humanity which it is his duty to show towards mankind [*sic*]. If he is not to stifle his human feelings, he must practice kindness towards animals, for he [*sic*] who is cruel to animals becomes hard also in his [*sic*] dealings with men [*sic*]. (Kant, 1963)

Thus, Kant argued that the kindness that humans show to other species is an indicator for human morality but not within the domain of morality and felt that humans should treat other animals with kindness in order to preserve their own humanity.

Similarly, Plous (2003) considered "historic attempts to portray human targets of prejudice as animal-like" (p. 510) as an attempt to use speciesism (a term for prejudice against non-human animals) to justify human oppression. Plous (2003) wrote,

> African Americans have been depicted as apes, Jews as vermin, women as prey, homosexuals as beasts, fat people as cows and pigs. Yet the very act of "treating people like animals" would lose its meaning if animals were treated well. Just as racism, sexism, and other prejudices share a similar mindset, many of the psychological factors that underlie speciesism serve to reinforce and promote prejudice against humans. These factors include power, privilege, dominance, control, entitlement, and the need to reduce cognitive dissonance when committing harmful acts. (p. 510)

Thus, Plous (2003) not only saw the treatment of non-human animals as an indicator for human morality, but also saw a complete parallel between the mistreatment of humans and the mistreatment of other species. Additionally, Plous (2003) held that animals all have intrinsic value and that treating animals better benefits all animals (including humans).

These considerations of the humane treatment of non-humans are an important consideration in social justice work. In societies where the mistreatment of other animals is normalized, arguments for human liberation often dismiss or condone the subjugation of non-human animals. For example, Freire's (1970/2000) position, which is hugely influential in social justice work, was that oppression of humans denies both oppressor and oppressed their humanness and reduces them to the status of animals. However, Freire did not challenge the status of animals and instead attempted to distance humans from other animals. Thus, by condoning the mistreatment of animals and distancing humans from other animals, the structures of oppression including hierarchy and domination remain unchallenged and undermine efforts for human liberation.[2] As I suggested after reviewing Bookchin's (1982) considerations of the relationship between subjugation of women and subjugation of Earth: any expression of oppression will spread like a virus throughout humanity.

In summary, not only is consideration of animals vital in understanding the structural influence of oppression on human moral considerations, but also caring for animals contributes to the liberation of all animals (humans included). By focusing explicitly on the rights of a narrow community, such as humans, we imply that other communities don't deserve rights and we debate over who should be allowed membership in that community. It may serve the welfare of human and non-human animals to instead extend moral consideration to other animals and establish universally consented upon rights to all animals.[3] Many of the rights that are associated with humanness should be extended to non-human animals, such as the right to live freely, the right to basic survival needs, and the right to live at all.

A final consideration at the interface of animal rights and conservation is zoos. Zoos are largely justified by their role in promoting "species conservation in natural habitats" (Myers, Saunders, & Garrett, 2004, p. 559). Zoos provide opportunity for humans to experience non-human animals in simulated natural habitats and such experiences may foster empathy for non-human animals and lead to greater concern for animals and the natural world. However, in a study on children's conceptions of the needs of animals, Myers et al. (2004) noted that "few subjects explicitly acknowledged freedom to roam as an important need, when it may be a precondition for meeting all other needs in an appropriate habitat" (p. 560). Myers et al. (2004) concluded that while data suggested that such value for freedom is related to the developmental stage of participants, the significance of normalization of the captivity of animals in urban environments could not be determined. Thus, zoos may have positive and negative influence regarding the subjugation of animals (including humans).

I maintain that the struggle against oppression requires uncompromising opposition of all oppression. I oppose the subjugation of all animals and the normalization of animals in captivity potentially perpetuated by many zoos. However, this has more to do with the treatment and portrayal of animals in captivity than it does with holding animals in captivity. Many beautiful animal sanctuaries provide the best possible conditions for animals who because of domestication, capture, or abuse cannot be safely reintroduced into the wild. Many of the facilities offer limited public access and facilitate educational experiences with constant emphasis on the welfare of the residents.

Cultural Normative Limitations

The Nazis used the term *Llebensunweten Lebens*, meaning "life unworthy of life" (Opotow, 1990). "Nazi propaganda defined Jews as outside the pale of humanity" (Oliner & Oliner, 1988, p. 6). Such condemnation and exclusion of the Jewish people was systematically normalized and enforced by the Nazis. This caused Hoffman (2000) to declare:

> Since the Holocaust, cultural relativism is dead. We no longer have the luxury of assuming every culture's values or guiding principles will pass the moral test and that each good is as good as any other. (p. 273)

It seems self-evident in consideration of the Holocaust that certain atrocities are universally and unquestionably wrong. However, cultural relativism may not be so easily dismissed. This section explores the influence of culture in an attempt to understand how moral injustice could occur with majority complicity on the scale of a society. The section begins by examining moral relativism and ethical objectivism. Next, the influence of cultural norms on individual and group moral judgments is examined while entertaining the possibility that such perceived injustice is objectively immoral and the possibility that morality is entirely subjective. Finally, I conclude with a pragmatic approach to morality in search of a middle ground between relativism and universalism.

Before continuing, I want to consider the framing of the Holocaust as the epitome of moral failure because of relativism. First, it stretches the concepts of moral and cultural relativism because German culture wasn't the culprit. What the Nazis (not the Germans) attempted was to establish a different epistemology—a competing reality wherein any dissenter was deemed unworthy of life. The Nazi party was no more representative of German culture than it was of dominant Western culture. If cultural relativism is denounced by citing the Holocaust, then the denouncement is of the moral validity of judgments relative to dominant Western culture. However, many of the liberating forces were also part of dominant Western culture. Thus, cultural relativism does not account for the Holocaust.

A second consideration of framing of the Holocaust as the epitome of moral failure is that Earth has not witnessed a single Holocaust, but many holocausts occurring before and after World War II. Defining the Holocaust as the single, most profound atrocity in the history of Western culture dismisses a history of atrocity perpetrated and perpetuated by dominant Western culture. If we are to suggest that moral relativism be denounced based on extreme cases, why not cite colonialism or slavery? I suggest that it is because the moral deficiency is easily recognizable in those cases as a deficiency of European and European American culture and in the case of the Holocaust, it is easy to maintain the moral standing of European American liberators. The pursuit of moral universalism must overcome ethnocentrism and egocentrism that underlie assertions of objectivity and lead to judgments of superiority.

Still, Hoffman (2000) raises an important question regarding the possibility of ethical objectivism. Regardless of the above considerations, most people, despite the subjectivity of our own experiences and differences in culture, maintain similar basic ideals about what is right and wrong. Such concepts were considered in the preceding discussion of rights. Despite the subjectivity of our individual experiences and the influence of normative forces such as culture, there is some innate intersubjectivity among all humans. This intersubjectivity allows us to cooperate and generally resolve moral conflicts constructively. When significant harm is perceived as the result of judgments made by majority consensus of a society, cultural relativism and ethical objectivism become important considerations.

Cultural or moral relativism suggests that morality is subjective and differs in accordance with culture. The presumptions of relativism are the following: (1) we are not able to know the moral truth of our judgments apart from our own culture, and (2) our moral judgments do not apply to those who do not share the same moral truths (Cook, 1999). Thus, moral behavior is dictated by cultural norms and judgments are only universally right or wrong in the context of a particular culture (e.g., in Hindu culture it is wrong to …). Moral relativism taken to the extreme suggests that all moral judgments are arbitrary and no moral claim could be better than any other. Ethical objectivism makes the opposite claims, suggesting that we can and do know moral truth and that it does apply to those who do not share the same convictions. The consequentialist position on moral objectivism is that right and wrong are determined by the outcomes of moral judgments and rely on a judgment of the desirability of those outcomes. The deontological position is that there are universal principles which rationally dictate whether or not a judgment is morally right or wrong (as illustrated by Kahn's [1999] definition of obligatory moral judgments).

Regardless of whether one believes that cultural norms validate moral judgments, there is no escaping the fact that culture is influential in individual moral judgments. Certainly the influence is not deterministic and individuals have the ability to resist or comply with cultural norms. We should never assume cultures are homogeneous or static or that all individuals approve of all of the practices of their culture. Martin Luther King Jr. expressed this sentiment beautifully when addressing the American Psychological Association in 1966.

> There are some things in our society, some things in our world, to which we should never be adjusted. There are some things to which we must always be maladjusted if we are to be people of good will. (as cited in Turiel, 2005, p. 4)

King's statement may represent a consequentialist perspective on the moral truths underlying the civil rights movement. In fact, throughout American history, cultural critique unapologetically condemning the outcomes of cultural practices has propelled our social progress (e.g., civil rights movement). King made it clear that the injustice he spoke of was just as wrong elsewhere in the world as it was in the United States.

The relativistic rejoinder to such a powerful stance on justice may suggest that the identification of injustice was relative to the status quo.

> It is easy for us to feel appalled at the way nobles exploited their serfs, plantation owners exploited their slaves and male chauvinists exploited women. But were these land-owners, slaveholders, and male chauvinists fundamentally different from us, or were they simply responding to different pressures and a different *status quo*? The prevailing power balances, then, seem to affect even the most aloof reformers' conceptions of social justice. (Austin & Hatfield, 1980, p. 53)

The above passage does not suggest that it is arbitrary to feel appalled by past injustice, but that we must understand the context of injustice. We have to be sympathetic but not permissive when critically examining the cultural influence of behavior that we consider harmful to ourselves or others.

When behavior is culturally normalized, individuals may not recognize their compliance as a moral judgment. The cultural normativity of the behavior may additionally be seen as validation of the moral judgment to comply. Therefore, we must critically examine our own enculturation. If we seek moral objectivity, we must consider how Western norms have been conducive to Western atrocity. Rather than denouncing the moral judgments of others, we must accept that Westerners may not be the moral light of the world (we don't have the greatest track record). However, although Western culture dominates Western society, Western society is not culturally homogeneous. Thus, we must look inward to critique the normative forces of our own cultures and societies and look outward, not in judgment, but in humility. As Grande (2004) wrote,

> Dominant patterns of belief and practice are…integrally related to the cultural and ecological crises. Thus, the need for understanding other cultural patterns as legitimate and competing sources of knowledge is critical. (p. 65)

However, Grande (2004) did not assert that colonized peoples exist beyond critique and referred to the idea as "not only short sighted but also patronizing" (p. 84). She clarified that

> voices of Indigenous and other non-Western peoples become increasingly vital, not because such peoples categorically possess any kind of magical, mystical power to fix countless generations of abuse and neglect, but because non-Western peoples and nations exist as living critiques of the dominant culture, providing critique-al knowledge and potentially transformative paradigms. (Grande, 2004, p. 65)

Thus, through exploring relativism we may find moral truth. However, there may be multiple moral truths and truth may not be static or absolute. We must be firmly rooted in our own identities and concepts of morality while simultaneously respecting, honoring, and learning from differing practices in other cultures.

In summary, humans experience life subjectively and our perceptions, experiences, and identities influence our moral judgments. At issue is whether moral judgments can be right or wrong. I suggest that we know certain universal rights and wrongs, but others most certainly will continue to emerge and evolve over time. Genocide, slavery, murder, rape, and mutilation are examples of things that are unquestionably wrong. In less extreme cases, moral judgments are much more subjective and objective rights may be unknown or nonexistent.

Pragmatic ethics establish a reasonable middle ground between objective moral judgments and subjectivity based on consequentialism. Sayre's (1991) environmental ethics fit well into my broad conception of the environment as natural and social. Sayre proposed

> an alternative view according to which the aim of environmental ethics is (1) a clear understanding of how moral norms actually come to be instituted in a given society, (2) the analysis of the practical effect of such norms from an environmental perspective, and (3) an examination of the relative desirability of alternative norms in light of their environmental effects. (p. 195)

All that remains is to sort out what is desirable. Although it is certainly a point of subjectivity, we all have to choose a point where we stand firm in our own moral convictions on what is good and what isn't. For example, I believe that all life is worthy of life. In my own self-conception, the value that I place on life transcends concepts of cultural and moral relativism.

Sayre's (1991) ethics most importantly call for understanding. In all but the most extreme cases, moral judgments can be identified as harmful without condemning the moral judge(s) as individuals or a cultural group. When behavior that is accepted/normalized is harmful, I believe we have a duty as humans to intervene. However, intervention cannot be an imperialistic effort at coercion, which is both a violation of rights and an ineffective means of changing behavior. We have to heed Sayre's (1991) advice and recognize how and why the harmful behavior may be a function of cultural and moral norms. We must meet people where they are, help them expand their perspectives, see the consequences of their actions, and voluntarily come to new moral positions. However, we must also be prepared to expand our own perspectives and recognize the consequences of our own actions. I'm reminded of the narrative of a Christian missionary who lived with Pirahã Indians in the Amazon rainforest with the goal of learning their language in order to convert them but wound up not converting anyone (Everett, 2008). Instead, Everett (2008) was personally transformed by the experience and gained new respect for the Pirahã way of knowing and being entirely in the present moment. The lesson is that we need to know people before we make assumptions about their judgments or behavior.

In conclusion, I do not reject the possibility of moral objectivity, but I make no claims that humans know such objectivity in the complexity of human social relations. I am unwilling to quickly condemn cultural or individual practices as immoral and I hold that undesirable behavior isn't always immoral because judgments involve subjectivity. Consequences of judgments, however, are less subjective, and when one's moral judgments cause harm, intervention is necessary. In the most extreme circumstances, such as violent assault, considering the possibility that the transgressor is acting morally, gently intervening, and having a conversation about the transgression would be unreasonable. In such cases, only the moral judgment of the person or people intervening could possibly dictate the appropriate response.

Social Death

Lisa Marie Cacho (2012) takes a structural approach to understanding (in)justice through the concept of social death. Social death is significantly related to moral exclusion and empathic bias. However, Cacho's method of exploring social death as a process of exclusion provides a more sociological than psychological perspective. Social death is the outcome of a process in which those deemed "dangerous, undeserving or unintelligible" are stripped of political legitimacy and moral credibility and denied the right even to demand rights. The socially dead are legally and morally "ineligible for personhood" (Cacho, 2012).

Criminalization is one of the methods of exclusion that results in social death. Criminalization is the variety of "ideological and material processes that turn some people into criminals by making it all but impossible for them to be law-abiding" (Cacho, 2012) because "certain vulnerable and impoverished populations and places of color" are targeted by policies of regulation and containment (p. 5). Cacho argued that because legal policy is generally accepted as

> both ethical and irreproachable, the act of law-breaking reflects poorly on a person's moral character. If following the law (legitimate or not) determines whether a person is moral or immoral, it is all but impossible for people assigned to certain status categories to represent themselves as moral and deserving. (p. 4)

Therefore, Cacho (2012) addressed legal, social, and moral exclusion singularly for the targets of criminalization and social death.

The targets of criminalization are inescapably prejudged as immoral and criminal and "deemed deserving of discipline and punishment but not worthy of protection. They are not merely excluded from legal protection but criminalized as always already the object and target of law, never its authors or addressees" (Cacho, 2012, p. 5). The progression in judgment from illegal to immoral and therefore undeserving of rights renders the targets ineligible for personhood. This "not only forecloses empathy but does so through producing people and places always already subject to a form of discrimination believed to be both legitimate and deserved" (Cacho, 2012, p. 82).

Cacho (2012) is suggesting that laws (as practically synonymous with morals) are written against specific racialized identities, making not only someone's behavior but also their very existence illegal.

> People subjected to laws based on their (il)legal status—"illegal aliens," "gang members," "terrorist suspects"—are unable to comply with the "rule of law" because US law targets their beings and their bodies, not their behavior. They are denied not only the illusion of authorship but even the possibility of compliance. (Cacho, 2012, p. 6)

Cacho (2012) supported this proposition by analyzing media portrayals and legal proceedings related to immigration, terrorism, and gang violence and showed that status and identity dictate the way law targets individuals. One example of the way that identity and criminality are deeply intertwined was provided by the aftermath of Hurricane Katrina. Cacho (2012) juxtaposed images in the media of a Black man wading through floodwaters with food from a grocery store with a nearly identical image of White couple who had taken food from a grocery store. The Black man was portrayed as a *criminal* who *looted* property, while the White couple were portrayed as *victims* who *found* food in a grocery store. Cacho (2012) concluded that only a Black body makes certain crimes (looting) visible. More precisely, Black is criminalized as a de facto status crime.

> De facto status crimes can be defined as specific activities that are only transparently recognized as "criminal" when they are attached to statuses that invoke race (gang member), ethnicity ("illegal alien"), and/or national origin (suspected terrorist). (p. 43)

A compelling case that Cacho (2012) examined was the case of eight high school teenagers who planned and executed a targeted attack on Mexican migrant workers in July 2000. The teenagers explicitly sought to attack "illegal immigrants" (incidentally, none of the migrant workers were undocumented). They nearly beat a man to death and beat, robbed, and harassed several other migrant workers for three hours while shouting racial slurs. In the ensuing trial, the teenagers were portrayed as victims of gang enhancement charges (Proposition 21) that dictated that they should face enhanced sentencing and be tried as adults, charges that were clearly not intended for people like them. National media attention and public sympathy were invoked for these otherwise "good kids" who had just made a "mistake." As Cacho (2012) concluded, "The primary question regarding the assailants' case was whether their irrefutable guilt tainted their innate innocence" (p. 29). Their innate innocence was related to their status as affluent white U.S. American citizens. At the same time, many other Youth of Color received enhanced sentences under Proposition 21 with very little public media attention. The clear subtext is that the status of gang member is inseparable from race, status categories, and criminalized identities (Cacho, 2012).

Another component of social death is that not all identities are unquestionably protected and privileged. While some people/s are considered entitled to rights, others must be deemed deserving of rights.

> When a group's rights are presumed to be entitlements, discussions about injustice can focus on whether people's rights have been respected. But when a group's rights are not socially recognized, discussions revolve around whether or not the aggrieved group even deserves to speak out against discrimination or exploitation, which effectively subverts or forecloses any dialogue about the actual injustices. (Cacho, 2012, p. 139)

African Americans, in this light, are often portrayed as having "earned" rights (as citizens), while Latina/o migrant workers are portrayed as undeserving of rights (as immigrants). This rhetoric is used to put civil rights and immigrant rights at odds because civil rights are considered to have been earned while immigrant rights are not (Cacho, 2012). Cacho points out that it is the same institutions that render these folks socially dead that must be appealed to in order to establish deservingness of rights. An appeal for rights made within a system that functions to deny rights may be a case in which the master's tools, indeed, will never dismantle the master's house. Thus, deservingness "is an unachievable prerequisite for the conferral of rights and dignity that functions to align paid and unpaid workers with the regulating institutions and ideologies that keep them economically exploitable and legally vulnerable" (Cacho, 2012, p. 12).

The rights we need to be talking about are universal human rights. If rights are distributed based on deservingness, then even those who are morally included are only conditionally included and must continue to prove their deservingness and are always unprotected from moral exclusion and social death. Cacho (2012) suggested that we must "set aside the notion that rights are contingent (with citizenship as the prerequisite) and conditional (with deservingness as the proviso)" (p. 141). In summary,

> As criminal by being, unlawful by presence, and illegal by status, *[the socially dead] do not have the option to be law abiding*, which is always the absolute prerequisite for political rights legal recognition, and resource redistribution in the United States. When the subjugation is engendered, justified, and maintained by the law, legal recognition cannot be a permanent or meaningful solution to subjugation. Criminalization justifies people's ineligibility to personhood because it takes away the right to have rights. Consequentially, criminalization makes sense of the contradictions that ensue when according unequal access to legal universality. (Cacho, 2012, p. 8)

Finally, Cacho (2012) concluded with the value of hope:

> The space of social death is a desperate space, overwrought with and overdetermined by ideological contradictions of ineligible personhood.... And yet the space of social death is always graced with hope, courage and/or youthful idealism, where those who decide to take responsibility for the unprotected are always looking for and stepping on the pressure points that can barely manage the contradictions that their very being inspires. (Cacho, 2012, p. 145)

Outlying behavior

To be frank, I do not have the background or expertise to adequately explore differences in the moral development of individuals who are outliers in the process of empathy and moral development as I have described it. Hoffman (2000) suggested a potential recipe for psychopaths rooted in his moral development theory:

> Empathy potentially exists in everyone, it may be reduced by irritability, fearfulness, and other temperamental factors, and by depressive and autistic tendencies that interfere with mimicry, role-taking, and other empathy-arousing processes. Combining these temperamental factors with nonnurturant, excessively power-assertive life experiences may well produce individuals who cannot empathize (psychopaths?). (Hoffman, 2000, p. 282–283)

However, as an educator and an outlier, I hope to offer a sympathetic perspective and a passionate refusal to believe that anyone is fundamentally incapable of prosocial behavior.

Autism spectrum dis/order

The concept of "Autistic Psychopathy" was introduced by Hans Asperger in 1944 (Jones et al., 2010). People diagnosed with autism spectrum disorder may experience empathy differently, but there is no reason to suggest that autistic people lack the ability to care for others or to develop moral principles.

> The essential features of autism spectrum disorder are persistent impairment in reciprocal social communication and social interaction (Criterion A), and restricted, repetitive patterns of behavior, interests, or activities (Criterion B). These symptoms are present from early childhood and limit or impair everyday functioning (Criteria C and D). (American Psychiatric Association, 2013)

Criterion A raises questions about how autism spectrum disorder (ASD) relates to the empathy and moral development process. The *DSM-5* (2013) elaborated on Criterion A:

> A. Persistent deficits in social communication and social interaction across multiple contexts, as manifested by the following, currently or by history (examples are illustrative, not exhaustive; see text):
>
> 1. Deficits in social-emotional reciprocity, ranging, for example, from abnormal social approach and failure of normal back-and-forth conversation; to reduced sharing of interests, emotions, or affect; to failure to initiate or respond to social interactions.
> 2. Deficits in nonverbal communicative behaviors used for social interaction, ranging, for example, from poorly integrated verbal and nonverbal communication; to abnormalities in eye contact and body language or deficits in understanding and use of gestures; to a total lack of facial expressions and nonverbal communication.
> 3. Deficits in developing, maintaining, and understanding relationships, ranging, for example, from difficulties adjusting behavior to suit various social contexts; to difficulties in sharing imaginative play or in making friends; to absence of interest in peers. (American Psychiatric Association, 2013)

However, behavior by individuals with ASD that may be interpreted by neurotypical (NT) individuals as antisocial may not be done with malicious intent (Jones et al., 2010). Rather, such antisocial behavior could be explained as well-intentioned faux pas by "individuals who find the social world difficult to interpret" (Jones et al., 2010). Obviously, individuals with ASD, like all other individuals, may occasionally act with malice.

Individuals with ASD experience empathy in the affective sense similarly to NTs but have difficulty with more cognitive forms of empathy such as perspective taking (Jones et al., 2010).

> Individuals with ASD have difficulty in understanding the perspective of others and consequently may react in a seemingly cold and uncaring manner in real-life situations. However, if information is presented in a way that enables individuals with ASD to identify others' point of view, they appear to show as much concern and compassion as typically developing individuals. (Jones et al., 2010)

Co-occurrence of ASD and lack of affective empathy (commonly associated with psychopathy) is possible. Both conditions are believed to be highly heritable, but they are genetically independent. Thus, although both are sometimes considered empathy disorders, autism and psychopathy are significantly distinct from each other (Jones et al., 2010).

Although many folks on the autism spectrum are highly communicative, either verbally or non-verbally, ASD is primarily researched and described by neurotypical people. In *The Positive Psychology of Personal Transformation: Leveraging Resilience for Life Change*, Garbarino (2011), who has written more than twenty books related to childhood violence and trauma, boldly discusses his own autistic tendencies (it is not explicitly clear whether or not Garbarino has been diagnosed):

> They say that because of having trouble understanding the emotions of others a person with Asperger's[4] may be seen as egotistical, selfish, and uncaring. This is mostly not true of most of "us." I know it is not true of me. Let me make it clear that my intentions are good: I believe I am "a good person." At heart I am a person of light and spirit. I am kind, generous, compassionate, caring, and giving to the degree that I can "figure out" what the kind, generous, compassionate, caring, and giving thing to do would be. (p. 34)

Additionally, the book *Asperger Syndrome & Social Relationships: Adults Speak Out About Asperger Syndrome* (Edmonds & Beardon, 2008) provides a collection of essays about social relationships written by adults with Asperger Syndrome. The personal narratives reflect a diversity of unique people who connect with others in a variety of ways. Several of the contributors work in social professions such as health care, teaching, and social work and actively engage in helping others. Many of the essays certainly reflect difficulty with perspective taking, but many express the active use of induction (attempts to make generalizations about the effect of their actions on others). I was struck by how actively attentive the authors were to their own difficulty regarding issues such as perspective taking and theory of mind for the express purpose of helping other autistic people achieve greater success in social relations. With the attention, deliberation, and detail of any social scientist these folks deconstruct everyday social interactions in order to understand social convention to which many NTs don't have to pay specific attention.

The following quote is meant as a joke and lacks empirical support, but it offers a shift in perspective regarding conceptions of what a disorder is:

> Neurotypical syndrome is a neurobiological disorder characterized by preoccupation with social concerns, delusions of superiority, and obsession with conformity. Neurotypical individuals often assume that their experience of the world is either the only one, or the only correct one. NTs find it difficult to be alone. NTs are often intolerant of seemingly minor differences in others. When in groups NTs are socially and behaviorally rigid, and frequently insist upon the performance of dysfunctional, destructive, and even impossible rituals as a way of maintaining group identity. NTs find it difficult to communicate directly, and have a much higher incidence of lying as compared to persons on the autistic spectrum (Institute of the Study of the Neurologically Typical, http://isnt.autistics.org, accessed 11 February, 2008). (as cited in Edmonds & Beardon, 2008, p. 11)

I intentionally challenge the term "disorder." However, it is important to recognize that there are mental and physical disabilities as well as mental and physical different-abilities. By so frequently pathologizing children as disabled, we form our expectations for them based on perceived deficits rather than potential (Damon, 2009). Because "children are acutely aware of how they are being appraised, and they base their own self-conceptions largely on how others see them" (Damon, 2009, p. 169), we limit our children's capacities when we view them as disabled. Damon (2009) argued that "no child who retains a living brain is ever wholly or permanently disabled.... There are always useful and self-fulfilling things that a child can learn to do" (p. 171).

All people deserve an active role in the way that they self-identify and we all need a little extra help sometimes. Difference is not necessarily disability. However, those who need assistance in order to be healthy and successful in daily life should not confuse the politics of (dis)ability with their practical needs. Labeling a human being as high or low functioning based on that person's need for help is dehumanizing. But it is not dehumanizing to recognize and understand someone's need for help on an individual basis without presuming that the person's value is based on independence.

Psychopaths, sociopaths & antisocial behavior

"The essential feature of antisocial personality disorder is a pervasive pattern of disregard for, and violation of, the rights of others that begins in childhood or early adolescence and continues into adulthood" (American Psychiatric Association, 2013). Those diagnosed with antisocial personality disorder frequently lack empathy; however, a lack of empathy is not required for diagnosis (American Psychiatric Association, 2013). This definition neatly applies to the majority of repeat offenders in prisons. In fact, the prevalence of the disorder is above 70% in prisons and substance clinics (American Psychiatric Association, 2013). Antisocial personality disorder is also associated with low socioeconomic status and is more frequently seen in urban areas (American Psychiatric Association, 2013).

In everyday use, the labels "psychopath" and "sociopath" are used interchangeably and evoke images of a merciless criminal or murderer. The *DSM-5* does not use "psychopath" and "sociopath," and such conditions would be diagnosed as antisocial personality disorder (American Psychiatric Association, 2013). Pemment (2013) overviewed evidence for the biological nature of psychopathy and its relation to early childhood development and advocated for specific use of the terms "psychopath" and "sociopath." Pemment (2013) suggested that the psychopath does not feel empathy and thus has no sense of morality or simply no regard for moral convention because of neurological differences. Sociopaths are defined as having a sense of morality that does not match their cultural context. Sociopaths' belief that their actions are just modulates their feelings of empathy for a potential victim (Pemment, 2013). Although not caring about others is a possible outcome of the lack of empathy, a lack of empathy doesn't doom someone to antisocial behavior.

Furthermore, the terms "psychopath" and "sociopath" and associated empathic "deficiency" are generally only ascribed post hoc to those who engage in antisocial behavior. However, Haidt (2006) pointed out that folks who "lose most of their emotional lives" as a result of damage to the orbitofrontal cortex (the lower third of the frontal cortex) "perform normally on tests of intelligence and knowledge of social rules and moral principles" (p. 12). Thus, diminished emotional capacity is not intrinsically associated with antisocial behavior. The deficit that these folks most significantly suffer is indecision because without the immediate emotional response to a situation, the individual must logically examine all of the pros and cons of each possible choice and may find little reason to pick one or the other (Haidt,

2006). However, damage to the entire frontal cortex has caused people to become sexually aggressive and uninhibited (Haidt, 2006). Certainly, differences in the brain affect our behavior, but rarely do they dictate our behavior. Cultural influence is too complex and profound a factor for any behavior to be biologically determined. The general consensus is that some people are born with neurological deficits that are conducive to psychopathy but that environmental factors more significantly influence the manifestation of psychopathic behavior (Garbarino, Bradshaw, & Vorrasi, 2002).

If someone is already condemned, criminalized, and diagnosed with antisocial personality disorder or as a psychopath or sociopath, even that individual's innocence is presumed to be evidence of guilt. The person is presumed to be fully and entirely antisocial. Thus, what for a non-criminalized individual would be normal behavior is perceived as a cover for evildoings. Anyone considered a psychopath is easily rendered socially dead and morally excluded. In fact, the Bush administration employed a similar rhetoric to the descriptions of psychopaths in moral research in order to justify the invasion of Iraq and Afghanistan.

> The Bush administration constructed the Middle East as an area where "the condition" of "evildoing" was hidden and latent, posing permanent and unpredictable threats to civilization. Afghanistan and Iraq were constructed as terrorist "hideouts" or places with "hidden" weapons of mass destruction, and this presumption of guilt assumed the existence of unseen evidence that may not, in fact, have existed. In other words, Iraq and Afghanistan would be presumed guilty even if proven innocent because the requirements for exoneration were also the terms of indictment. (Cacho, 2012, p. 100)

Psychopathy, sociopathy, and antisocial personality disorder are easy answers for antisocial behavior (cf. Haidt, 2006, on the myth of pure evil) because they simply condemn the evil other as incapable of goodness. Evil may exist, and I won't discount the possibility that some contingent of the human population is simply "bad." However, it is too simple to suggest that people who engage in antisocial behavior feel no empathy and are simply compelled by their evil nature to commit crimes. We need to consider the possibility that something is very wrong in environments where antisocial behavior is prevalent.

Childhood bullying may be a warning sign for psychopathy in adulthood (Gini, Pozzoli, & Hauser, 2011). Gini et al. (2011) found that children who bully their peers often make sound moral judgments but lack moral compassion (i.e., psychopathy: they know right from wrong but don't care about the moral distinction). The study stressed the necessity of intervention in bullying behavior. However, a strictly behaviorist approach to antisocial behavior is not sufficient because "antisocial behavior in children appears to be correlated with physical abuse, parental conflict, and antisocial parents (among a number of other factors)" (Pemment, 2013, p. 2). Interventions must address the child's social environment and the process of moral development in addition to behavioral interventions.

Garbarino (1996) suggested that factors such as violence and economic poverty, especially in urban environments, are significant causes of antisocial behavior. Garbarino (1996) used the term "socially toxic environments" to describe the manner in which such circumstances are conducive to antisocial behavior:

> Social toxicity refers to the extent to which the social environment is psychologically poisonous, in the sense that it contains serious threats to the development of identity, competence, moral reasoning, trust, hope, and the other features of personality and ideology that make for success in school, family, work, and the community.... Like physical toxicity, social toxicity can be fatal in the form of suicide, homicide, drug-related and other life style–related preventable deaths. But mostly, it results in diminished "humanity" in the lives of children and youth by virtue of leading them to live in a state of degradation, whether they know it or not. (Garbarino, 2011, p. 30)

Furthermore, the *DSM-5* stated, "Concerns have been raised that the diagnosis may at times be misapplied to individuals in settings in which seemingly antisocial behavior may be part of a protective survival strategy" (American Psychiatric Association, 2013). It should be considered that if such a survival strategy was persistently necessary throughout a child's development, "survival strategy" in a toxic environment could be incorporated into one's moral framework and become disorder. The motivation for antisocial behavior and associated "lack of remorse" characterizing psychopaths may be reflective of moral and empathic limitations that all humans are subject to such as habituation.

There may be two unique processes at work: (1) antisocial behavior is a product of toxic environments, and diminished or altered empathy and morality allow one to cope with having acted in such a way; (2) antisocial behavior is a result of diminished or altered empathy and morality, and the prevalence of such behavior produces toxic environments. To ask which is more prevalent is the wrong question. Both are at work and together the two processes make up a positive feedback cycle that creates and reproduces toxic environments (social degradation) and perpetuates antisocial behavior.

Hunger and malnutrition also contribute to antisocial behavior. Many vitamins, minerals, and amino acids are vital to cognitive functioning and affect production of chemicals in the brain, such as serotonin, dopamine, and norepinephrine (Massachusetts General Hospital, 2006). Studies have shown that dietary correction or supplementation can reduce antisocial behavior and violent tendencies in incarcerated adults (Gesch, Hammond, Hampson, Eves, & Crowder, 2002) and adolescents (Schoenthaler et al., 1997), and improve cognitive functioning (Rogers, 2001) and academic performance (Florence, 2007). In the Schoenthaler et al. (1997) study, the researchers collected and analyzed blood samples prior to and during the intervention study. The subjects who had deficiencies (most common deficiencies were the vitamins B_1, B_3, B_6, and B_9, all associated with mood disorders) prior to the study displayed the most significant decrease in violent behavior. Rogers (2001) concluded that "dietary risk factors for physical (ill)health are also often risk factors for mental (ill)health" (2001). Ultimately, a healthy diet and healthy brain improve cognitive functioning and may help people resolve conflicts nonviolently.

When survival is at stake, in cases such as hunger, the pursuit of nourishment is one's primary motivator. Considerations will likely become egoistic as a matter of survival and one may find behavior otherwise considered immoral to be justifiable or may act without moral consideration. Additionally, antisocial behavior in the face of injustice may be an act of symbolic defiance. Maybe we shouldn't simply consider the "rightness" of the struggling individual's behavior. Indeed, maybe we should consider that if someone has to resort to criminal(ized) behavior to acquire food, regardless of the individual's personal merit, there may be a significant moral shortcoming on the part of all of those complicit with a system allowing anyone to go hungry.

Just as unmet physiological needs may encourage a tendency toward antisocial behavior, so too do unmet psychological needs. Antisocial behavior that individuals may justify based on their personal needs is not always morally excusable even in the context of unmet needs, environmental stressors, and pathology. I draw the line not at the loss of property but at the infliction of physical harm: it is excusable to steal food if one is starving, but it is not excusable to shoot a convenience store clerk to acquire food. In these extremely antisocial cases, individuals often make little distinction between survival needs and desires (Gibbs, 2014). One such case of atrocity resulting from unmet psychological needs for love and acceptance is the recent murder spree committed by twenty-two-year-old Elliot Rodger. In videos posted on YouTube under the title *Life Is So Unfair Because Girls Don't Like Me* (Rodger, 2014), Elliot blamed women for his unfulfilled need for love and acceptance. He later murdered six students who attended his university (Pengelly, 2014). Elliot was obviously mentally ill and took deeply misogynistic elements of society to an atrocious extreme. However, despite all pathology, and how "sick" Elliot might have been, there is no possibility of suggesting that Elliot's unmet needs excused his actions. Pathologizing behavior does not justify it or relieve those who have done such severe harm of responsibility. However, we need to be looking more carefully at social structural influences on antisocial behavior.

In summary, we can't immediately dismiss people who do bad things as bad people. Reductionist theories are always appealing, but human behavior is not that simple. To dismiss antisocial behavior lets those of us who have self-conceptions as good people off the hook too easily. The idea that we are responsible only for our own actions is nonsense; we co-construct our environment together and we (all people) co-construct environments that are conducive to antisocial behavior. We must work to change the environments conducive to the behavior and to rehabilitate the individuals who have acted antisocially (cf. Chapter 8, "Treating Antisocial Behavior," in Gibbs, 2014). Hope mandates that even in the most deterministic cases (e.g., psychopaths), we critically consider how, when, and where we can make the most effective interventions in any process that produces undesirable outcomes. This means that regardless of biological influence (which educators have little ability to intervene in, but psychiatrists may) we must make social and environmental interventions. We need to create environments and foster experiences that are more conducive to prosocial behavior than to antisocial behavior.

Trauma & chronic stress

Severe trauma can stimulate psychopathy (Garbarino, 2011). Traumatic events are typically defined as confrontations involving the actuality or threat of death or serious injury (Alisic et al., 2011). "Traumatic events may disrupt a youth's brain development, developmental skills, talents, personality development, and functioning" (Nader, 2011). Nader suggested that less violent events such as prolonged separation for young children or shame and humiliation for older children can also be traumatic.

> Psychological distress following exposure to a traumatic or stressful event is quite variable. In some cases, symptoms can be well understood within an anxiety- or fear-based context. It is clear, however, that many individuals who have been exposed to a traumatic or stressful event exhibit a phenotype in which, rather than anxiety- or fear-based symptoms, the most prominent clinical characteristics are anhedonic [inability to experience pleasure in normally pleasurable acts] and dysphoric symptoms, externalizing angry and aggressive symptoms, or dissociative symptoms. (American Psychiatric Association, 2013)

Garbarino (2011) described trauma as simultaneous affective and cognitive over-arousal. The overwhelming negative feelings are coupled with thoughts "beyond normal ideas of human reality" (Garbarino, 2011, p. 49).

Posttraumatic growth, resilience, and coping

"The experience of a single incident of acute trauma in an otherwise good life is likely to yield to short-term rest and therapy…that brings a traumatized person's feelings and thoughts back into normal focus" (Garbarino, 2011, p. 50). Human response to trauma is complex and may, but usually doesn't, result in posttraumatic stress disorder (Alisic et al., 2011). It may have transitory or lasting ill effects but it can also promote emotional growth. Children can recover and even grow from traumatic one-time events when their environment is otherwise positive, but a single event, and especially chronic stress or trauma, can negatively affect the development of a child's brain (Haidt, 2006). The role of a trusted adult (e.g., parent/guardian or teacher) is significant in providing assurance and a safe and consistent environment (e.g., home, school). Children need to know that "their world" is OK (physical and emotional safety and stability) even if the outside world is not (Garbarino, 2011).

Posttraumatic growth suggests that

> making sense of the bad things can lead to some good things—if you are fortunate in who your friends are, what resources you have, how your temperament equips you to deal with stress and threat, and how you come out on the other side with a new appreciation for the meaning of life. (Garbarino, 2011, p. 47)

Haidt (2006) identified three types of posttraumatic growth. The first suggests that by getting through a traumatic experience and the resulting emotions, one can discover a sense of inner strength and emerge with a new understanding of their own abilities that positively changes self-concept. This knowledge of the ability to cope with difficult situations can be a source of strength in future traumatic experiences. Haidt (2006) refers to this newfound ability to cope as something akin to inoculation against future trauma. The second type of posttraumatic growth that Haidt (2006) discussed can have a filtering effect on personal relationships. In the aftermath of trauma, one's most important relationships sometimes become more apparent. One's friends either step up to help or step back, and the emotional vulnerability and dependence on friends and loved ones during recovery from trauma can strengthen those relationships. Ultimately, people who undergo profound challenges together may emerge closer to one another than ever before or, conversely, be pushed apart. Third, trauma shakes the foundation of our reality and may thus provide a stimulus for individuals to reassess their philosophy and priorities in life. I think we all can point to examples of these three types of posttraumatic growth such as a close friendship developing through a shared hardship or people who recover from cancer and resolve to quit their office job to pursue their dreams.

People's ability to maintain hope in the face of trauma and achieve growth is referred to as resilience. Resilience can be considered as functional and existential (Garbarino, 2011). Functional resilience may allow someone to continue a normal routine in the face of trauma, but they might still be suffering inside, and trauma symptoms could surface years later (Garbarino, 2011). The aim of positive psychology is not just surviving, but thriving (functional and existential resilience). Garbarino (2011) described a human quality called hardiness that protects against traumatization. Hardiness is composed of a positive outlook on life, a sense of control, and a disposition to see challenges rather than threats.

Finally, Garbarino (2011) most significantly discussed the importance of forgiveness and compassion. Violence is cyclical and the victim of one trauma may become the perpetrator of future trauma. For inner peace and peace in the social environment, forgiveness is vital and "we must not allow our hurt egos and our dark sides to use the opportunity presented by traumatic events to liberate and validate our rage" (Garbarino, 2011, p. 60). We must also be compassionate, in that regard, to the possibility that those who have caused our own trauma have done so as a result of their own trauma. This is about healing, breaking the cycle of trauma, not about excusing or validating the transgression. Through compassion and forgiveness, some violent youths that Garbarino (2011) worked with

> have undergone a profound spiritual transformation as part and parcel of fully owning their crimes and simultaneously seeing the roots of that crime in the way they themselves were victimized and traumatized by others, earlier in their lives. (p. 60)

Thus, injustice and trauma offer opportunity for spiritual growth but not without love, hope, and support.

Just because positive psychology helps us to see positive outcomes of trauma doesn't make trauma good. As Haidt (2006) stated:

> I don't want to celebrate suffering, prescribe trauma to everyone, or minimize the moral imperative to reduce it where we can.... I want only to make the point that suffering is not always all bad for all people. There is usually some good mixed in with the bad, and those who find it have found something precious: a key to moral and spiritual development. (p. 141)

The potential for growth doesn't justify one's suffering. We need to work to reduce trauma, but we must focus on positive outcomes when trauma does occur. In our pursuit to stop the degradation of ourselves and of the world, we will never eliminate all opportunities for suffering. Adversity is a natural part of life, and certain potentially traumatic events such as death can never be prevented.

Chronic trauma

The negative psychological effects of trauma are not always transient. Adequate support and physical and emotional safety necessary to recuperate may not be attainable or an individual may be differently equipped to cope. It may also be possible that some traumatic experiences may simply be too much for anyone to cope with. Acceptance of such reality is challenging, and positive psychology is important for both reading about trauma[5] and treating it. As Garbarino (2011) said, making sense of bad things can lead to good things. Thus, the harmful outcomes of trauma must also be addressed with the hope that they can be shifted toward the positive.

Posttraumatic stress disorder (PTSD)

> is characterized by overwhelming feelings of reexperiencing the traumatic event (e.g., nightmares and intrusive thoughts), by the avoidance of stimuli and emotional numbing (e.g., avoiding places related to the event and feeling detached from others), and by symptoms of hyperarousal (e.g., concentration difficulties and hypervigilance). (Alisic et al., 2011)

Nietlisbach et al. (2010) found that affective empathy was diminished by PTSD but cognitive empathy was not. The participants did not have difficulty understanding the emotional states of others, but were not emotionally affected and this was explained by dissociation. The dissociation and diminished feelings of affect were speculated to be a coping strategy to prevent being overwhelmed by others' emotions (Nietlisbach et al., 2010). Certainly this coping strategy associated with traumatization is consistent with Hoffman's (2000) discussion of empathic over-arousal. Such dissociation is also consistent with psychopathy and antisocial personality disorder. The environmental correlates with PTSD are also similar to those of APD, including lower socioeconomic status, lower education, childhood adversity, and minority racial/ethnic status (American Psychiatric Association, 2013). Indeed, Martens (2005) suggested that childhood trauma promotes the development of antisocial personality disorder and, in turn, antisocial personality disorder places an individual at greater risk for "violent assaultive traumatization as a consequence of reckless, impulsive, self-destructive, and violent attitudes and social-emotional and moral incapacities, and retraumatization" (Martens, 2005).

Violence is prevalent in U.S. American culture and communities and so are symptoms of trauma. It is not uncommon for students to arrive at school on a Monday morning having lost a peer, friend, or family member to violence over the weekend. Violence is perpetrated horizontally (within our social groups, e.g., gang violence), vertically (by those with greater agency/power, e.g., police involved in shootings), and on ourselves (e.g., drugs and suicide). Such is the reality that many youth face and this violence and trauma may be significantly influencing their development (Cloitre et al., 2009).

> Furthermore, youth living in violent communities may experience "pathological adaptations" such as hopelessness, fatalistic thoughts, desensitization to violence, and truncated moral development. These youth often participate in high risk behaviors such as alcohol or drug abuse, promiscuous sex, or association with dangerous people. (Garbarino et al., 2002)

Violence spreads through communities like a virus. High-violence, high-economic-poverty urban environments (however, rural environments are not immune) suffer a vicious cycle wherein the victims of violence become perpetrators of further violence.[6] Such environments, as Garbarino (1996) said, are toxic and they are traumatizing our youths. However, psychopaths and sociopaths thrive in socially toxic environments (Garbarino, 2011).

When trauma is a pervasive occurrence in the lives of children and adolescents, the psychological effects may differ significantly. Chronic trauma occurs

> when "normal" is the problem, not the solution. Chronic traumatic danger imposes a requirement for what we might call "developmental adjustment," in the sense that it demands a reorganizing of one's very understanding of life over time and not just a relatively short period of recalibration. (Garbarino, 2011, p. 50)

Chronic trauma has also been described as complex or persistent PTSD.

> Exposure to sustained, repeated or multiple traumas, particularly in the childhood years, has been proposed to result in a complex symptom presentation that includes not only posttraumatic stress symptoms, but also other symptoms reflecting disturbances predominantly in affective and interpersonal self-regulatory capacities such as difficulties with anxious arousal, anger management, dissociative symptoms, and aggressive or socially avoidant behaviors. (Cloitre et al., 2009)

Gabarino (2011) suggested that individuals (and groups) faced with chronic trauma developmentally adjust to accommodate a reality that threatens their sense of reality. When individuals persistently face atrocity in their life their brain adjusts in order to accommodate that reality (Garbarino, 2011). This can potentially result in the compromise of an individual's basic instinct for survival because the reality of danger is normalized.

However, it's not quite that simple. Aside from those who may be somehow predisposed to a lack of empathy that may be exacerbated by chronic trauma, hypersensitivity to future trauma is a more common outcome (Garbarino, 2011). Garbarino (2011) suggested that although it may seem like one's affect has completely disappeared, seeming lack of emotion may be symptomatic of chronic trauma that has "not yet reached the point of collapse" (Garbarino, 2011). This distinction is particularly relevant to criminal (in)justice because many youths who had committed murder or other violent crimes that Garbarino (2011) interviewed were believed to be suffering from chronic trauma. Garbarino (2011) suggested that the "most violent youths construct elaborate defense mechanisms against anxiety, fear, and abandonment; these defense mechanisms culminate in the persona of the cold-blooded 'gangster'" (Garbarino, 2011, p. 56). Thus, an apparent lack of remorse for a violent crime may not be an actual lack of remorse.

Cacho (2012) problematized how "criminalized poor of color are characterized as either products of violent environments that should be heavily policed or as irrational people incapable of moral agency who need to be under police surveillance" (p. 83). Overemphasis on structure and overemphasis on one's own agency both condemn victims of trauma and targets of oppression. Overemphasis on structure makes situations seem entirely determined and overemphasis on agency leads to victim blaming. Those affected by trauma are not doomed. They are not on a one-way road to further violence, psychopathic behavior, prison, drugs, and death. Garbarino (2011) refers to this as terminal thinking. It is hope's kryptonite.

Stress

It is not only when the emotional injury that we inflict on our youths results in trauma that concern is warranted. Institutional racism and other forms of oppression such as economic poverty are detrimental to psychological and physical health.

> Poorer communities are less likely to have adequate health and social services, creating a problem of access and timely use. Also, the physical environments are more likely to expose the residents to health hazards (e.g., air pollution, lead, dust, dirt, smog, and other hazardous conditions). Finally, the concentration of poverty and its related characteristics (e.g., exposure to drugs, crime, gangs, and violence; unemployment, stress, and anxiety; substandard housing and schools; and lack of green space or fresh fruits and vegetables) often creates social environments that lessen social connectedness and provide fewer social benefits for residents. (Mays, Cochran, & Barnes, 2007)

"In racially segregated, poor neighborhoods, both chronic and acute daily stressors (e.g., violence, unemployment, personal safety concerns) repeatedly invoke a biological challenge similar to the flight/fight response" (Mays et al., 2007). The accumulation of chronic stress results in "premature wearing down of the body," referred to as allostatic load, and it is associated with heart disease, chronic inflammation, and cognitive impairment (Mays et al., 2007). We all know through our own experience that stress also affects our self-care. Rogers (2001) speculated that because carbohydrate-rich junk foods can increase serotonin in the brain, carb cravings may be our body's way of self-medicating for depression. However, in cases in which the stressors are persistent, such self-medication can cause diet-related health problems such as obesity and diabetes.

The ill effects of chronic stress are not exclusively correlated with economic poverty or economic poverty as a result racial discrimination. Chronic stress is associated with racial discrimination independent of socioeconomic status. People experiencing race-based discrimination have physiological responses including "elevated blood pressure and heart rate, production of biochemical reactions and hypervigilance" that eventually result in disease and mortality (Mays et al., 2007). African Americans are disproportionately affected by diabetes, cardiovascular heart disease, hypertension, and obesity and the disparity is not attributable to genetics or socioeconomic status (Mays et al., 2007). African American women, regardless of socioeconomic status, suffer the highest rates of preterm birth and their babies weigh the least (on average). The health disparities affecting African Americans are posited to relate to "cultural differences in lifestyle patterns, inherited risks, and social inequalities that are reflected in discrepancies in access to health care, variations in health providers' behaviors, differences in socioeconomic position, and residential segregation" (Mays et al., 2007). Although there is no biological validity to the concept of race (Laland & Brown, 2011), the biological results of the social construct of race are real (Mays et al., 2007).

This paradoxically calls into question notions of race-based medicine, which essentially reduces to racial profiling. It is dangerous to make medical assumptions based on perceptions of racial experience. It's especially scary when drugs are prescribed exclusively to African Americans (e.g., BiDil). However, we do see that the social construct of race has real consequences for health. I propose that the only just race-based medicine is social justice. However, self-medication in the form of alcohol, tobacco, and illicit drug use is also correlated with stress associated with racial discrimination and internalized racism—likely as a coping strategy (Mays et al., 2007). However, such drug use often exacerbates stress and increases health problems. Additionally, illicit drugs are often associated with violence, increasing the likelihood that one will experience trauma.

As in the case of trauma, chronic stress causes youths' response to stress to become erratic, resulting in more frequent involvement in violent situations as victim or perpetrator (Duncan-Andrade, 2010). Again, it is important to note that neither trauma nor chronic stress dictates that any individual will behave antisocially. We (all people) need to take responsibility individually and collectively for the structural influences on stress and trauma and there are many levels of collective membership that must be considered. Family groups, blocks, neighborhoods, schools, workplaces, communities, cities, states, nations, humans, and members of the community of life all have important roles to play in reducing human violence.

I refuse to be the devil's advocate by entertaining the possibility that anyone is suffering and dying because they are bad, unworthy of rights, and ultimately deserving of pain, suffering, and death. I maintain that such violence is not inherent; it is inflicted. We inflict trauma on ourselves; we inflict it horizontally to family and peers in similar status groups; and we inflict it vertically on those less protected and less privileged than we are. The underlying moral shortcoming is not just on the part of the repeat offenders, the gangs, or drug dealers (as criminal or criminalized people). We are all co-responsible. We need to work to reduce violence, trauma, and chronic stress by addressing the injustices that cause them. We also need to help those who are experiencing trauma to cope with their situation. Education can play a fundamental role in raising consciousness to help students work for social change and reduced violence. However, there are two practical actions that schools need to be taking: (1) provide basic mental health training to all teachers so that they may recognize trauma, and (2) provide mental health services.

Catalyzing Action

From the most significant actions, such as risking one's own life to save the life of a stranger, to the mundane acts of kindness and generosity that humans perform on a daily basis and go largely unnoticed, humans act altruistically. Oliner and Oliner (1988) suggest that at the point at which a person decides whether or not to act, the decision is already largely made because that person has already internalized the systems for response. Intuitive or instantaneous decisions likely result from previous experience and previous contemplation. Haidt (2006) suggested that

> moral judgment is like aesthetic judgment. When you see a painting, you usually know instantly and automatically whether you like it. If someone asks you to explain your judgment you confabulate. You don't really know why you think something is beautiful, but your interpreter module (the rider) is skilled at making up reasons.[7] (p. 21)

Haidt (2006) didn't argue that we never know our motivations, but that we do not always explicitly understand our motivations. Sometimes we contemplate, make inductions, and come to new conclusions and sometimes we act upon the inductions and principles we've already constructed and internalized.

Intuitive moral and aesthetic judgments may be associated with natural tendencies, but are certainly related to previous experiences. The more I study, the more I see a deep relationship between aesthetics and morality. I think that the path to beauty is the path to justice and vice versa. The gut feelings we so often (at least I so often) act upon are exactly this combination of esthetic and internalization. Consider the following lines from "The Road Not Taken" by Robert Frost:

> Two roads diverged in a yellow wood,
> And sorry I could not travel both
> And be one traveler, long I stood
> And looked down one as far as I could
> To where it bent in the undergrowth;
>
> Then took the other, as just as fair
> And having perhaps the better claim,
> Because it was grassy and wanted wear;
> Though as for that the passing there
> Had worn them really about the same

Frost ultimately made a gut decision and tentatively tried to explain why, but admits that the reason doesn't really hold up—confabulation. In the process of this research there has been much that I have lain awake at night contemplating, trying to understand, and trying to understand how what I've learned fits within my existing moral framework. I've most significantly wondered how I act upon my new understandings. However, in my moral judgments, despite all my contemplation, I've ultimately followed my gut. A more substantial amount of what I've learned and written were things I either already knew tacitly without knowing explicitly or deeply and tacitly realized that I knew as I read. Much of this work consists of truisms explored empirically, and clichés elaborated and contemplated upon.

The above considerations are central to understanding how and why we take action when we are confronted with a moral dilemma, especially when action is not explicitly demanded of us. Before considering more explicitly how people are morally motivated, I want to consider an example from my own experience researching morality. Throughout the course of working on this book, I've wondered whether my own behavior has been affected by the knowledge I have attained. Has researching morality made me more sensitive to moral conflicts or better equipped to help others recognize and resolve moral conflicts positively? I've often answered no, because when I am with young people I am entirely in the moment and I am not thinking about theory. Although I occasionally do step back from the experience and attempt to consider how the topics of this book apply, I have never found myself considering theory when attempting to help kids resolve a conflict.

Oliner and Oliner (1988) are right to suggest that the experiences culminating in a triggering or catalyzing event more significantly indicate how someone will respond to that event than the event itself. However, the nature of the event is also significant and will be considered further in the context of a profoundly salient catalyzing event—the Holocaust. At the end of this chapter, I explore how such triggering may apply in situations in which the moral conflict is less monumental (e.g., climate change). Triggering or catalyzing events challenge the morals to which one has already committed (Hoffman, 2000). For some Holocaust rescuers, inaction threatened their self-conception to a point at which they didn't consider their behavior to be a choice (Oliner & Oliner, 1988). They felt they had to act because they felt that they could not have lived with themselves if they hadn't. Some felt others' suffering as if it were their own. Others felt that they had simply reacted to the circumstances that they encountered without significant reflection. A common sentiment among rescuers was, "I did nothing unusual; anyone would have done the same thing in my place" (Oliner & Oliner, 1988, p. 113), a remarkably humble, but inaccurate sentiment.

Oliner and Oliner (1988) identify four circumstances that facilitated action: information and comprehension of the need, risk, resources, and a precipitating occasion. The information available to rescuers and their comprehension of the events were not significantly different from that of non-rescuers. Information alone was necessary, but not sufficient for action. Similarly, the perceived risks for rescuers and non-rescuers were not significantly different. The only difference was that rescuers acted despite that risk. In terms of material resources, rich and poor alike rose to the occasion, but the resource that did significantly differ between rescuers and non-rescuers was community. Rescuers more frequently belonged to organizations or had personal relations that they believed would support their efforts. Still this was not exclusively a characteristic of rescuers: some acted alone, keeping their activities secret from friends, family, and neighbors even beyond the end of the war. Many of those rescuers stated that their interview was the first time that they had discussed their activities. Finally, the precipitating occasion was the most significant indicator for rescue. More rescuers were asked for help than non-rescuers. This could be due, in part, to the rescuers, through past actions, relationships, and character traits, seeming more likely to help. This relates to the concept of extensive persons. Still, others acted without being asked. These circumstances all facilitated rescue in certain ways, but it is clear that altruistic behavior is not dictated by circumstance alone.

> Researchers found that the actions of the individuals they interviewed were based only partially on their hope that they could create change. Often these individuals acted in spite of not knowing whether they would actually make a difference. These activists, at times, "decide that success is not important—that something must be done even though there is no assurance that it will change anything." For them, the essential "rightness" of the action makes the issue of success irrelevant. (Berman, 1997, pp. 46–47)

The nature of the precipitating occasion that catalyzes action varies. However, it seems universal that some event, perceived or experienced, triggers altruistic behavior. Oliner and Oliner (1988) describe three types of catalysts: those that aroused or heightened the observer's empathy, those that challenged a normative demand, and those that violated the observer's principles (Oliner & Oliner, 1988). While Oliner and Oliner (1988) recognize that some rescuers deliberately sought out ways to help, most did so in response to a specific event that they witnessed. Even those who sought out opportunities to help without having directly witnessed any of the atrocities of the Holocaust were certainly acting in response to the immediacy and severity of the situation. Still, it was more than simply the situation that dictated responses from the rescuers; their actions reflected their individual character. Such a quality of their character is illustrated by the fact that the majority of the interviewed rescuers continued to be actively engaged in helping behavior beyond the war (Oliner & Oliner, 1988).

When faced with such a precipitous occasion the moral limitations and empathic bias that have been described all influence whether or not someone will act. Additionally, the ways that individuals process information and cope psychologically are influential. Oliner and Oliner (1988) wrote that "interpretations of events are human inventions, and that what and how we choose to see shape our responses—and thus the future" (p. 260). Humans have an emotional preference for knowledge that we want to be true and avoid thinking about things we don't want to be true (Clayton & Myers, 2009). This bias may be especially pronounced when it comes to beliefs about our own identities. "Defensive thinking results when our basic wants, such as the desire for comfort and pleasure, are incompatible with our rational or moral judgment" (Clayton & Myers, 2009, p. 26). Such denial may be a very significant barrier to catalyzing action in conflicts from the most salient (such as the Holocaust) to the least visible (such as climate change). In fact, some non-rescuers, interviewed by Oliner and Oliner (1988), reported that they simply couldn't believe that something so unthinkable was happening.

Learned hopelessness may result from exposure to events that seem uncontrollable and as a potential outcome of inaction in the face of precipitous occasions (Christens, Collura, & Tahir, 2013). Learned hopelessness may further diminish moral action in future conflicts. Some rescuers, however, acted despite very low feelings of self-efficacy (Oliner & Oliner, 1988). Some rescuers even continued their efforts despite witnessing the murders of those they had risked their lives to help (Oliner & Oliner, 1988). However, many would be unlikely to take action, especially at such great personal risk, if they felt that their efforts would be unsuccessful. Furthermore, a sense of self-efficacy, despite not necessarily precluding moral action in conflicts that we perceive (Oliner & Oliner, 1988), may affect the way we perceive of those conflicts (Clayton & Myers, 2009).

In the spaces of privilege, the effects of the socio-ecological crisis are not always obvious.

> Faced with a conflict between a desire for self-gratification through unsustainable behavior and the knowledge that the environment is threatened by such behavior, we repress our awareness of the conflict, deny the threats that face us, displace them onto other communities, and rationalize our continued unsustainable behavior as having no alternative. (Clayton & Myers, 2009, p. 26)

Clayton and Myers (2009) explicitly refer to ecological threats; however, the same process certainly applies to humanitarian and social justice concerns that are inseparable from ecological issues. Such defensive thinking is more prevalent when one does not feel able to resolve conflicts between personal needs and moral principles (Clayton & Opotow, 2003). Johns (2013) wrote that "conservationists…resist taking on structural change because it seems to require that people make material sacrifices, and this is unrealistic. But is giving up mindlessly complex social hierarchies that infantilize people a sacrifice?" (p. 238). Thus, Johns (2013) resists framing environmentalism as conflict between consumerism and sustainability and frames the conflict as one between consumerism (and more generally capitalism) and human welfare (and more generally planetary welfare). In Johns (2013), framing pro-environmental behavior is also prosocial and the struggle is against unjust social structure that affects consumerism. However, Clayton and Myers (2009) frame the conflict in such a way that our own agency in sustainable behavior is primary. How we perceive conflicts certainly influences the ways in which we are motivated to action. It may be that in some cases or for some individuals framing a conflict in a certain way may be more effective than framing it in some other way.

All humans develop moral principles that guide our determinations of what is fair and acceptable behavior. We are often unknowingly complicit in injustice and maintain a self-conception that mandates justice. In such cases, triggering events do not necessarily need to be external. If through conscientization we begin to understand the ways in which our own actions contribute to injustice, continuation of such behavior would be a violation of our established moral principles. It would be collusion rather than unintentional complicity. bell hooks (2003), an artist, teacher, and notable feminist,[8] suggested that people choose to collude or oppose racism, but I think that people need experience, opportunity, and agency to recognize it as a decision to oppose or collude.

Collusion with injustice would threaten the self-conception of anyone who maintains justice as a moral mandate.

> People affirm their sense of self by endorsing moral mandates consistent with their personal values. Moral mandates are the selective expression of moral values that are central to people's sense of identity. (Clayton & Opotow, 2003, p. 310)

This violation of an individual's moral mandate threatens one's sense of identity. The cognitive dissonance that such a violation incites can be resolved with a change in behavior or through justification and restructuring of moral principles as a defense. Despite the fact that humans have a natural tendency or motivation to cooperate and help one another, we do not always help. When relations of power are involved, moral mandates could conflict with the status quo and a behavior change could seem difficult or impossible. In such a situation one could succumb to learned hopelessness (Christens et al., 2013). However, hopefulness—beneficial effects of perceived or actual control of one's circumstance—can also be learned (Christens et al., 2013) and must be addressed through pedagogy.

The value of risk as a catalyst honors "the fact that we may learn and grow in circumstances where we do not feel safe" (hooks, 2003, p. 64). Thus, hooks (2003) concluded that "the presence of conflict is not necessarily negative but rather its meaning is determined by how we cope with that conflict" (p. 64). Conflict should be considered as a process and an outcome. Conflict processes can be positive (such as cooperation) or negative (such as disrespect) and lead to constructive and destructive conflict outcomes (Opotow & Weiss, 2000). In cases related to social and ecological degradation, where the issues may not always be salient, a conflict may be necessary in order to catalyze moral action.

In more recent years, many of the victims of international policies and trade agreements such as NAFTA and CAFTA have been catalyzed to action around the world by a perceived threat not only to their traditions and sovereignty but to their very existence. Some refer to this as the fourth world war (where World War III is defined as the Cold War). The Zapatistas, for example, felt that NAFTA and the dissolution of communally held property eliminated their means for subsistence. Proclaiming that NAFTA was a death sentence, the Zapatistas essentially seceded from Mexico (Hayden, 2002). External threats can increase solidarity and catalyze collective action. When folks have little left to lose and everything to gain and perceive inaction as a death sentence, fighting injustice isn't an altruistic motivation but a survival instinct. In some parts of the world, and almost certainly increasingly in more "developed" nations, the ecological crisis is becoming such a catalyst. However, a significant motivation for this work is my hope that some of us will be catalyzed to action before we are confronted with death.[9]

In summary, a precipitous occasion so pronounced as the Holocaust is not necessary for action. The three types of catalysts (those that aroused or heightened the observer's empathy, those that challenged a normative demand, and those that violated the observer's principles) also provide insight for motivation to action for social and ecological justice. In the case of environmental action, the same considerations of information and comprehension of need, risk, resources, and a precipitating occasion apply. The need for environmental action is not always so apparent. For example, the effects of climate change are not necessarily apparent in all parts of the world to the untrained eye. Education is vital, then, to make the issues salient and to catalyze action.

Conclusion: It Takes All Types

Oliner and Oliner (1988) identified four characteristic types of rescuers based on their personalities and motivations. Although the rescuers could not all be neatly categorized into a single type, these idealized types illuminate certain characteristics that may promote altruistic behavior. The types illustrate that all sorts of people rise to challenging occasions and that there is no universal set of circumstances, experiences, or characteristics that cause a person to be a hero or to simply act in accordance with principles that are considered prosocial. As educators, I believe it is our role to recognize and support the development of moral principles of all people, and never to devalue the different ways in which others are prosocially motivated.

The four types represent rescuers who were primarily empathically motivated, normatively motivated, motivated by a principle of justice, and motivated by a principle of care. In each case, rescuers were rarely exclusively driven by one of the four archetypes, but instead by some combination of them. It was also recognized that the rescuer's motives changed over time. Still, by understanding the four motives as archetypes, the differences among rescuers are illuminated.

Thirty-seven percent of rescuers were initially empathically motivated and acted primarily in response to the empathy they felt for the individuals whom they helped. Their efforts and attention were primarily focused on the individuals whom they aided and they more commonly felt close personal attachment to those whom they helped. Oliner and Oliner (1988) discuss one rescuer, called Stanislaus, in detail. As a prototypically empathically motivated rescuer, "understanding others, taking their perspective, and anticipating their futures may have left Stanislaus little psychological room to consider his own needs" (p. 197). Interestingly, Stanislaus was reported to feel a strong external locus of control and lower-than-average self-esteem. Oliner and Oliner (1988) describe Stanislaus as having a "fatalistic orientation towards life." Interestingly, Stanislaus doesn't fit neatly into the category of extensive or constricted persons.[10] Stanislaus's accounts seldom focus on himself and describe the victims and their situations in detail. A sense of connection to others and empathy guided Stanislaus's actions.

Fifty-two percent of rescuers were initially primarily normatively motivated. They were not typically motivated by their connection to the victim. They were motivated by a sense that implicit or explicit societal rules that dictate normal activity and behavior were being broken. These norms sometimes related to a church, national identity, or other social institution where a rescuer was directly influenced by the inspiration of a leader or action based on internalized norms. Normatively motivated rescuers often felt a sense of obligation or duty to help and anticipated feelings of guilt over inaction. They often viewed themselves as part of a larger cause. For example, rescue was a matter of national pride for many Danish people. The sense of normlessness and loss of order that the war provoked threatened the reality in which these people lived, and promoted behavior that accorded with the rules of decency to which they ascribed before the war and fostered a return to normalcy.

Principle-based motivation accounted for 11% of rescuers' initial action. This includes those motivated by principles of justice and care. These people were inspired to act because what was happening violated their moral principles and threatened their sense of self. These rescuers were often motivated to act despite any perceived futility of their actions because their actions were in line with their principles. Those motivated by justice tended to have more impersonal relationships with those that they rescued but expressed feelings of anger or hatred toward the perpetrators of injustice. These people more frequently acted independently. A rescuer called Suzanne, who was highly motivated by principle, expressed her actions as simply cognitively logical. However, as Hoffman (2000) suggests, this sort of principled response is often "hot cognition." Suzanne stated that

> it was evident that a dictatorship had begun. I knew that one of the first measures would be an indictment against the Jews. I did not react to the first indictment, but then the second statute was published, I decided to get involved. I wrote a letter to the three rabbis in my region, and as I remember what I wrote, it said:

Sirs:

> I am very upset that in my country, in the twentieth century, some citizens are persecuted for their religious or racial conditions. My ancestors, Protestants of Cevennes, have fought for their freedom of belief. I cannot but follow their example and at this time, I will be at your side. Can you put me in touch with some needy French families belonging to your faith so that I may be of some help. (Oliner & Oliner, 1988, pp. 210–211)

From this point forward, Suzanne actively sought out ways to help and acted primarily independently. Her descriptions provide accounts of the sequence of events with few to no descriptions of individuals or discussion of feelings. Suzanne is reported to have helped several hundred people, all of whom were strangers to her and who had not asked her for help. Suzanne's behavior may relate to the empathic limitations discussed earlier. Suzanne was considered to be much less empathic than average (two standards of deviation below) and so she was less disposed to empathic over-arousal.

Rescuers motivated primarily by the principle of care felt "a dominating sense of obligation to help all people out of a spirit of generosity and concern for their welfare" (Oliner & Oliner, 1988, p. 217). A rescuer called Louisa expressed concern for all people to a point at which she put herself, her husband, and her son in positions of significantly increased risk for the benefit of the larger group. In her accounts, she sometimes did not distinguish between her own family and those whom she worked to rescue. Louisa stated in her interview,

> My mother said, "I don't think you have the right to do this. Your responsibility is for the safety of your own children." I said to her that it was more important for our children to have parents who have done what they felt they had to do, even if it costs us our lives.[11] It will be better for them—even if we don't make it. They will know we did what we felt we had to do. This is better than if we think first of our safety. (Oliner & Oliner, 1988, p. 217)

Although these are all accounts of Holocaust rescuers, Hoffman (2000) and Berman (1997) use these and other accounts to generalize to situations of similar gravity and to the everyday activities in our personal lives.

In summary, human behavior is the result of complex interactions between internal and external processes throughout one's development. Part II barely scraped the surface of human motivation and morality. However, the lesson is that we would all do well to be active practitioners of empathy and sympathy, both through our behavior and in our judgment of others' behavior. Pedagogy must always leave space for all people and all voices to develop in their own unique ways. This is not to say that it is not the job of educators to create space and foster experience that is conducive to moral development or specific principles. A lesson from Freire and Freire (1992/1994) concludes Part II, "Empathy and Morality," and transitions to Part III, "Pedagogy for Restoration."

> What kind of educator would I be if I did not feel moved by a powerful impulse to seek, without lying, convincing arguments in defense of the dreams for which I struggle, in defense of the "why" of the hope with which I act as an educator? (p. 83)

Freire and Freire (1992/1994) expressed this question rhetorically, but continued to state that:

> What is not permissible to be doing is to conceal truths, deny information, impose principles, eviscerate the educands for their freedom, or punish them, no matter by what method, if, for various reasons, they fail to accept my discourse—reject my utopia. (p. 83)

Freire and Freire (1992/1994) beautifully argue that education is not value neutral and this book reflects that sentiment. I maintain, as a matter of faith, that every living human, from the deviant to the exemplar, has a greater potential for "goodness" than they have realized. My entire exploration of morality is guided by my faith in humanity, in all of life, and in Earth. An important illustration of this point is my investigation into outlying behavior. To my knowledge, no portion of that exploration is inaccurate, but it is certainly not complete. I refuse to take myself or the reader down a path of hopelessness.

PART III: PEDAGOGY FOR RESTORATION

> Dear Teacher:
>
> I am a survivor of a concentration camp. My eyes saw what no man [*sic*] should witness: Gas chambers built by learned engineers. Infants killed by trained nurses. Women and babies shot and burned by high school and college graduates.
>
> So, I am suspicious of education. My request is: Help your students become human. Your efforts must never produce learned monsters, skilled psychopaths, educated Eichmanns. Reading, writing, and arithmetic are important only if they serve to make our children more human.
>
> —Author unknown (as cited in Ginott, 1972, p. 87)

What follows is my attempt to conceptualize pedagogy for restoration. This final part of the book significantly departs from the preceding attempts to understand the nature of oppression and human morality. Pedagogy for restoration is not pedagogy in reaction to oppression. No set of instructions for liberation or restoration could ever be prescribed. Thus, pedagogy for restoration maintains focus on youth-potential rather than ideology. The goal is to conceptualize an approach that would help young people reach a physical, emotional, and intellectual place where they feel able to follow their own passions, their own sense of right and wrong, and their own vision for a better world. Pedagogy for restoration is a process of healing on the individual level, collectively, and for the planet. I believe such healing or self-restoration could spread through social systems just as I have suggested that oppression does. Such efforts by educators in all walks of life promote the global project of restoration.

Restoration must be a process of healing personal, ecological, and social degradation. The degradation of the natural world caused by humans and the degradation of humans caused by humans are not inevitably unidirectional processes. As I have argued, the planet can and will heal; I see restoration as a process allowing humans to heal with the planet. Our presence on this Earth has not been a single consistent expression of degradation.

Introduction to Pedagogy

> Being a responsible member of the global community demands…action to promote a more just, peaceful, and ecologically sound world through the appreciation of diversity, respect for human rights, and the peaceful resolution of conflict. (Berman, 1997, p. 190)

I see pedagogy as something beyond the science, art, and theory of teaching. My own pedagogy is deeply concerned with the philosophy of education, how we relate with our students, why we teach, and the goals of education beyond curricular objectives—to create a better society. I have studied, combined, and attempted to reimagine the work of many teachers and intellectuals rather than extensively reviewing their works. Notably among the works that I have studied are those of Paulo Freire. Freire's vision for critical consciousness, of reading the word and reading the world, of the necessity of hope and the pursuit of utopia are also my vision. Although it is not my intention to essentialize Freire or to suggest that his works are beyond critique, much of what he wrote applies just as much today, if not more, than when his works were published. However, Freire (1992/1994) held that the meaning of his books (or any book) was not found through passively reading the words, but through critically engaging with the ideas and making one's own meanings. Freire's intention was for his work to be critiqued and reimagined for application in different contexts rather than wholly adopted. I hold the same intentions for my work. I do hope that my pedagogy (or any portion of this project) could be useful to others, but just as I will be adapting and reimagining this work for the rest of my life, anyone who finds value in these pages must do the same.

Critical Pedagogy

Critical pedagogy brings a socially conscious lens to education that aims to reflexively critique all of educational and social practice. Critical pedagogy must fundamentally be an action-oriented anti-oppression approach to education. There is enough in the world that needs critique that we may omit many critiques without damage. Critique solely as an exercise in deconstructionism is of no utility to critical pedagogy or an anti-oppression effort. My primary critique of the body of work of critical pedagogy in the decades following *Pedagogy of the Oppressed* (Freire, 1970/2000) is the overemphasis on critique over creativity.

Weiner (2007) suggested that critical pedagogy suffers from a lack of clear definition as well as an overemphasis on disciplining itself within particular ideologies (e.g., Marxism). Weiner (2007) further suggested that critical pedagogy is in the middle of a crisis of imagination and suggested that we "put the goal of free and imaginative thinking above the goal of ideological persuasion" (p. 68) to "create context for rejuvenated imagination and its development" (p. 67). Debates over the necessity of Marxist language, an explicitly socialist agenda, and of the centrality of class struggle become detrimental to the struggle against oppression and the project of restoration when they overshadow the real work of making change. As the Zapatistas say, we need to be working for a world where many worlds fit. I'm not interested in a debate over specific ideology. Rather than a coercive approach that aims to liberate students from capitalism only to indoctrinate them in Marxism[1] (or any ideology), we need to aim only to facilitate experiences that broaden perspectives, encourage critical thinking, and create a space that promotes self-liberation. I firmly believe that it is not our business, as educators, to enforce any particular way of thinking. To deny my own political agenda would be disingenuous, but to assert that there is only room for my own ideology in a project of restoration would be disparaging to the project of restoration.[2]

The stakes are too high to quibble over Marxism. Some critical authors and educators take such a hard line on particular theories that creative thinking is overwhelmed by a demand for ideological consistency. Karl Marx famously addressed such ideological stubbornness by proclaiming: "what is certain is that I myself am not a Marxist" (marxists.org). Marx made this statement during a disagreement over the *Programme of the (French) Workers' Party* (1880). Marx believed that Jules Guesde (the leader of the Workers' Party) was too dismissive of the value of reformist workers' struggles and too strictly insisted on a radical-revolutionary interpretation of Marxism (marxists.org).

Thus, Marx made it absolutely clear that he would not allow his thinking to be marooned in his own ideology by denouncing Marxism. This was certainly a little hyperbolic, but made the point quite well. Critical pedagogy may be suffering a crisis of the critical imaginary, as Weiner (2007) suggested, because of strict adherence to ideology outlined by folks (e.g., Marx and Freire) who never intended such a strict and narrow interpretation of their own works. Wiener (2007) suggested that "even if it is true that everyday life is shaped by ideologies, it is less clear that solutions to troubles caused by ideology will be discovered in ideology" (p. 63).

However, Wiener's (2007) critique of critical pedagogy was not purely deconstructive. Weiner (2007) proposed four guiding principles for critical pedagogy to resolve the crisis: **(1) The primary concern must be ending oppression in all forms.** Weiner (2007) explicitly identified the difference between "a more complex understanding of how oppression works in all its forms" and "transforming conditions and relations of oppression" (p. 61). Weiner did not dismiss the former but emphasized the latter. **(2) This work needs to center those whom it hopes to help.** This refers to both the language that we use and the locations (primarily in the academy) where these conversations take place. Both language and location often exclude those who should be centered. Weiner (2007) wrote that the primary audience for critical pedagogy are privileged individuals who are not the targets of systematic oppression and reminded us that "it does not matter if what you say or write is brilliant if no one who you want to communicate with can understand a thing you say" (p. 60). **(3) The work must reach beyond its own ideology.** This refers to the importance of the critical imaginary and of not marooning ourselves or our educational practice in Marxist territory (or any ideology). **(4) There is a need for public intellectuals.** This refers to the necessity of prioritizing bringing theory into the real world rather than advancing complex theoretical research. People like Noam Chomsky, bell hooks, Subcomandante Insurgente Marcos, Paulo Freire, and Chuck D[3] are public intellectuals who have worked in various ways to bring critical consciousness to the public. Pedagogy for restoration rests on these theoretical foundations.

Moral Education

> What is required is nothing less than institutionalized structures that promote supportive relationships with the same seriousness as is currently devoted to academic achievement. (Oliner & Oliner, 1988, p. 259)

Most of education, at least the curricular objectives, and the focus of standardized tests concern explicit knowledge—facts and formulas—but the more important goal of understanding and meaning making is more difficult to define, discipline, and assess.

> If moral instruction imparts only *explicit* knowledge (facts that the rider can state), it will have no effect on the elephant, and therefore little effect on behavior. Moral education must also impart *tacit* knowledge—skills of social perception and social emotion so finely tuned that one automatically *feels* the right thing in each situation, *knows* the right thing to do, and then *wants* to do it. Morality, for the ancients, was a kind of practical wisdom. (Haidt, 2006, p. 160)

Tacit knowledge can be constructed through induction based on explicit knowledge and through experience. In education, the experience encompasses everything that happens in the classroom (or any educational space) from the most subtle to the most overt. The environment that we construct in education and the relationship between teacher and student are hugely influential. Students are always learning, but much of what they learn in school is not what is written into the curriculum. *The Moral Life of Schools* (Jackson, 1993) explored the ways in which the subtle or even mundane routines of classroom life may have profound moral implications. This encompasses everything from the cleanliness and organization of the classroom, to the decorations on the walls, to the tone of the teacher's voice.

As educators, we must be sure our students know three things: I hear you, I accept you, and I am with you no matter what. As the exploration of empathy has shown, we may never achieve a perfect match between our empathic feelings and our students' experience. We may never completely hear anyone (in the sense that we perceive their inner states). However, it is a powerful aspiration to seek to fully hear, accept, and support students. I believe this is the true work of education and if we do this work, the simple vocation of teaching, of sharing information on a path to understanding, critical consciousness, and enlightenment requires very little theory.[4]

Humans are moral by nature, so moral education needs only to provide opportunity for children to develop and commit to their own moral principles. This most significantly happens through action and reflection (e.g., transgression and induction). Moral development research shows that children need to make their own inductions (often with the guidance of an adult) rather than be told explicitly what is right and wrong (Hoffman, 2000). Therefore, moral education must be conceptualized as a path of moral growth and development with an emphasis on the walking (experience and moral action) rather than didactic instruction in morality.[5] Moral education must be about the real-world and lived experience rather than academic artifacts of the real world. Education must be about creating and protecting space for development, for free thinking, and for imagination. However, it can't be an entirely hands-off approach. The approach must be flexible and adaptive, but it cannot be entirely without structure or direction. "Maintaining intentionality and trajectory in the midst of real-world messiness is demanding, far more demanding than standing at the edge of the wilderness and giving the expeditionary team a lecture about what they might find there" (Palmer & Zajonc, 2010, p. 40). It's a journey we take together.

Pedagogy for restoration does not heavily focus on specific moral principles or an attempt to design moral curriculum but on walking with students and assisting them along the path. Rather than charting the course, educators help students learn to read the map, chart their own course, and take the lead. We provide experience, knowledge, and logistical support, but our greatest success is marked by our students' knowledge that their accomplishments are their own (Simpson & Hsih, 2003). However in all things, harmony is found in the balance between extremes. Students should recognize that their achievements are their own, but with the humility to recognize that others have helped them. This calls for balance between populism and elitism (and similarly authoritative and authoritarian approaches) (Freire & Freire, 1992/1994). A fully populist (entirely submissive to public opinion) approach would not provide the direction or perspective necessary to raise consciousness, but an elitist approach in which the educator presumes to know precisely what the students need to know would be little more than a pedantic and didactic attempt at indoctrination.

There can't be a scripted *how* for restorative education. The *how* has to come out of the student-teacher-community relationship. I have tried to focus on the *what* and the *why* with the presumption that a framework is tentatively established wherein each educator's own *how* would emerge. However, the fact that the approach cannot be canned and standardized does not imply that every educator needs to start entirely from scratch. Leading critical educators need to produce "concrete, hands-on examples[6] to provide educators with both inspiration and ideas for developing a parallel unit on a similar or different theme or to spin off an activity and add their own creative questions in a range of content areas" (Bode, 2012, p. 343). In the following discussions of pedagogy, I have tried to provide examples showing how such ideas are being put into practice in ways that are both academically rigorous and socially/ecologically important.

Pedagogy for restoration is composed of three elements: Pedagogy of basic needs, pedagogy of identity, and pedagogy of purpose. By addressing students' basic survival needs (see Chapter 6, "Pedagogy of Basic Needs"), we make the journey possible. By helping them to know themselves and in the process coming to know them deeply (see Chapter 7, "Pedagogy of Identity"), we begin to make the path as we walk it. And by searching for and constructing meaning and purpose (see Chapter 8, "Pedagogy of Purpose,"), we help students connect their sense of self with their community and the planet in such a way that they commit to the path.

· 6 ·

PEDAGOGY OF BASIC NEEDS

> [When someone is] chronically and extremely hungry...Utopia can be defined very simply as a place where there is plenty of food.... Life itself tends to be defined in terms of eating. Anything else will be defined as unimportant. Freedom, love, community feeling, respect, philosophy, may all be waved aside as fripperies which are useless since they fail to fill the stomach. (Maslow, 1943)

Making sure that students have their basic needs met, including emotional needs, must precede any educational effort. Not only do I know this tacitly, in my gut, to be morally imperative, it is explicitly and empirically true (as I will show). Therefore, education must become a healing process before it can become a learning process (cf. Duncan-Andrade, 2010, on education as a healing process). All educators need to think more deeply about our role as healers: to serve our students and to promote their well-being in any way we can. We must assist our students in reaching a place where they are physically and emotionally prepared to learn.

Educator-healers should look to the healers of their own cultures and world cultures for guidance. In environmental education, it is common for instructors to have wilderness medicine training because we often take students into the backcountry and need to be prepared to respond to a medical emergency. This training is not peripheral to education, but can, should, and does inform pedagogy. In wilderness medicine (and all emergency medicine), the same initial assessment is performed on every patient, whether they have a minor scrape or a sucking chest wound, in order to avoid overlooking symptoms of potentially life-threatening injuries. We assess symptoms in the order that they are most likely to kill our patient and we stop and treat any condition that we discover in the order that it is identified during the assessment (Tilton, 2010). There are countless examples of patients dying because the first responder was distracted by an obvious symptom while the patient died of a less obvious injury. The stop-and-fix assessment is designed to prevent such potentially deadly oversights.

Educators as healers need to be performing stop-and-fix educational assessments based on Maslow's (1943) hierarchy of needs.[1] Maslow suggested that humans are motivated by the desire to satisfy certain universal human needs. Starting with the most basic needs, one will generally only be motivated to satisfy each "higher" need once the lower needs are met (usually depicted as a pyramid or tipi). The most basic or "lowest" needs are physiological (e.g., food, water, shelter) and safety. Maslow suggested that these needs may entirely dominate one's consciousness if unmet and that the "higher" needs can only be fulfilled once the basic needs are sufficiently met. There are, however, notable and profound exceptions in which humans forgo their basic needs (e.g., Gandhi's hunger strikes). Psychological needs of belonging (e.g., love and acceptance) and esteem are next on the tipi. Maslow (1943) recognized that these needs may not be prevalent equally for all people (e.g., for some self-esteem may be of higher importance than love) and when unfulfilled, the "higher" are not as all-consuming as the basic needs. Finally, when one has satisfied one's basic and psychological needs, that person can seek self-fulfillment or self-actualization.

Unfulfilled physiological and safety needs are educationally (as well as morally) stop-and-fix situations. Hunger is a stop-and-fix. Trauma is a stop-and-fix. The stakes are not always as high in education as they are in first aid, but still we must consider how overlooking the physical and emotional well-being of our students can be catastrophic (academic failure and social or even physical death). Kids are arriving at school (or any formal or informal educational space) in conditions not conducive to their education or not arriving at all. Instructional methods and learning theory don't make any difference if students are hungry. Thus, pedagogy of basic needs focused on meeting children's physiological and safety needs is the first step in pedagogy for restoration.

Hunger may represent the most prevalent unmet and inadequately met physiological need that U.S. students face. "Food is essential to academic performance because it provides the energy necessary for cognition" (Woodhouse & Mark, 2012). For the purposes of this discussion, I intend a broad interpretation of hunger. Hunger does not refer strictly to starvation; it includes students who may have missed breakfast and/or lunch and so their attention suffers, but who otherwise may be well nourished. We all know what it's like to skip lunch and have trouble focusing on school or work later in the afternoon. However, fewer know what it's like to miss a meal every day, or several times per week, and so be unsure whether there will be enough to eat for the next meal and when that meal will be. This is what so many of our students experience: not starvation, but food insecurity. Food insecurity is much more prevalent in the United States than starvation.

Children who are not physically experiencing hunger pangs (contractions of the stomach due to severe hunger) may be deficient in any number of micronutrients necessary to keep their body functioning and developing optimally and their brain healthy and ready to learn. In the United States, children who are suffering health effects of malnutrition may, in fact, be obese. Childhood obesity is often associated with micronutrient deficiencies resulting from limited access to healthy foods (Drewnowski & Specter, 2004). Calorie-dense, nutrient-poor foods are cheaper, more convenient, and more easily available to folks living in economic poverty (Drewnowski & Specter, 2004). Overconsumption of macronutrients and underconsumption of micronutrients can lead to obesity and deficiencies (Drewnowski & Specter, 2004).

Malnourishment is not a simple issue of teaching folks to eat better. It is a moral issue. For example, by suggesting that obesity is always a sign of excess and poor education, we ignore structural and economic injustices and justify oppression by judging individuals as ignorant and gluttonous based on their body's appearance. Few would be so callous as to suggest that someone displaying more recognizable signs of malnutrition (e.g., emaciation and bloating) was simply lazy or ignorant of the necessity of a healthy diet. However, we constantly blame others for their ill health, which is commonly (at least partially) associated with structural oppression. Although food and health education are vitally important and overconsumption can be harmful to individual health and the health of the planet, we need to be considering how stress and limited time, resources, and access, as structural injustices, contribute to obesity and myriad diet-related health problems (e.g., heart disease and diabetes) and developmental disabilities prevalent in United States.

Many parents of malnourished children are not ignorant about how they should feed their kids, and I maintain that all parents want the best for their children. However, because of the injustice of poverty and corrupt economic systems, many do not have the time, access, and ability to feed their children healthier foods. The documentary *A Place at the Table* provides a look into the reality of hunger in United States, including several interviews with parents struggling to provide their children with both enough food and healthy food (Jacobson & Silverbush, 2012). Such narratives become experiential (even through film) and offer greater potential for empathic arousal than intellectual accounts of hunger. These narratives make issues real in a way that empiricism and statistics cannot. Seeing the pain in one mother's eyes as she discusses her children's malnourishment, the stress associated with struggling to get food on the table for dinner every night, and the agony of sending her children to bed hungry may be more moving than the knowledge that 17 million families in the United States (Weaver-Hightower, 2011) face food insecurity. Together, in dialectical solidarity, facts and the feeling, the reading of the word, and the reading of the world make this a moral issue and demand action.

Malnourished children have frequently been shown to have decreased attendance, attention, and academic performance along with various other health issues (Florence, 2007; Rogers, 2001; Woodhouse & Mark, 2012).

> The human brain needs sufficient energy—specifically glucose—and a variety of micronutrients to perform cognitive functions. A long-term deficiency of any or numerous macro- or micro-nutrients causes malnutrition and consequential cognitive impairment, the extent of which depends on the duration and degree of the malnourishment and the timing of its occurrence in development. In the United States, macronutrient malnutrition (i.e., starvation) is rare, but the diets of America's schoolchildren lack quality as measured by adequate and varied consumption of fruits, vegetables, and whole grains, and moderation of saturated fats and extra-calorie foods. Therefore, it can be inferred that US students' brains are often malnourished, as they are undersupplied of the micronutrients needed for effective cognition. (Woodhouse & Mark, 2012)

Numerous studies have found significant correlation between decreased overall diet quality and poor performance on academic assessments (Florence, 2007). However, as Florence (2007) discussed, income is a significant indicator for both malnourishment and academic success so it is often difficult to attribute causality to diet apart from income (and other socio-economic factors). Children are likely eating an inadequate diet, in part, because of their socio-economic status. The ill effects on their overall health and cognition influence their academic and economic attainment and contribute to generational poverty. However, such chicken-and-the-egg questions are less important for educator-healers than they are for scientists. We know that the brain needs sufficient macronutrients (calories) and micronutrients (vitamins and minerals) to function healthily and develop. We also know that economic poverty presents myriad barriers to educational opportunity and attainment. We cannot focus on either/or solutions (poverty or hunger); we must address both explicitly and simultaneously in whatever ways we can.

Feeding children breakfast has a significant impact on cognition, behavior, and performance (Florence, 2007). Simply stated, hungry students don't learn optimally. More precisely, these students don't optimally learn the lessons on the curriculum. As Maslow (1943) suggested, unmet basic needs, especially physiological and safety needs, become all consuming, and one can be motivated to achieve "higher" needs only when the "lower" needs are met. As Maslow (1943) wrote, when the need for food is unsatisfied,

> all capacities are put into the service of hunger-satisfaction, and the organization of these capacities is almost entirely determined by the one purpose of satisfying hunger. The receptors and effectors, the intelligence, memory, habits, all may now be defined simply as hunger-gratifying tools. Capacities that are not useful for this purpose lie dormant, or are pushed into the background. The urge to write poetry, the desire to acquire an automobile, the interest in American history, the desire for a new pair of shoes are, in the extreme case, forgotten or become of secondary importance. For the man [*sic*] who is extremely and dangerously hungry, no other interests exist but food. He dreams food, he remembers food, he [*sic*] thinks about food, he emotes only about food, he [*sic*] perceives only food and he [*sic*] wants only food.

If even one of our students is hungry, the first lesson of the day needs to be breakfast. No other lesson matters. If educators serve only to ensure that school (or any space) is a safe place where kids are served a healthy breakfast and lunch every day and sent home with a healthy snack, it would revolutionize education, even if kids spent the rest of the day on the playground. I hypothesize that happy, healthy kids would learn more on the playground than hungry kids in the classroom. However, there is tremendous educational opportunity in meeting the basic needs of students.

Maybe it's unfair to suggest that the role of education is to ensure that all students' basic needs are met, both inside and outside of school. However, education is part of a public institution whose concern is exactly that: the welfare of the public. If the larger institution doesn't meet its mandate, then education is undermined from the start. This leaves educators with three options. (1) We can keep doing more of the same. We can keep teaching the standards and attempting to prepare kids for standardized tests knowingly leaving many behind. (2) We can defer education and focus entirely on meeting basic needs. (3) We can take up the slack of failing and floundering institutions as an educational effort. Such an education in the real world, addressing hunger or other community needs, prepares students to be actively engaged citizens and provides context and meaning for traditional learning objectives.

It may go without saying that I demand option three. Food is not only prerequisite to learning; meals are opportunities to learn. Priority number one must be to get food into the mouths of hungry students. Second must be for the food to be healthy. Third is to sustain healthy eating. The remainder of this section is focused on attaining and sustaining food security through education. I explore how helping students and communities meet their basic needs can achieve moral and traditional education goals. However, constant attention to the physical reality and immediacy of hunger must be maintained. It would be extremely distasteful to ignore the immediate reality of a student's physical hunger while working with the class to address access and food-security in their community.

Shifting the Discourse

> Learning is often conceptualized as occurring in a social vacuum—the black box of research—denuded of concerns of the body, its needs, its pleasures, and its politics. Food, as highly body-centric, thus might seem unrelated to schooling's purest mission, the acquisition of skills and knowledge. (Weaver-Hightower, 2011)

Pedagogy of basic needs frames the fulfillment of physiological and safety needs as a matter of justice that demands action. In such a framing, the discourse on need is shifted: the shame and stigma associated with need (e.g., homelessness) is replaced with guilt[2] associated with privilege. However, it is not the right of an educator to design the discourse, but instead to join with the students and community to seek an understanding of the issue beyond cursory explanations that are so frequently accepted (e.g., justifications based on considerations of merit that lead to victim blaming). With understanding and contemplation the moral conflicts of the issue become more salient and inaction due to ignorance and lack of contemplation becomes informed collusion (e.g., realizing that one's advantage is unfair or comes at the expense of others). Although we have many psychological defenses, this conflict threatens our self-conception and demands resolution either by cognitively restructuring (e.g., defensive thinking) or taking action (see the section "Catalyzing Action" in Chapter 5).

Positive psychology provides some tools to prevent defensive thinking from interfering with positive resolution of moral conflicts (i.e., action). We need to know and to help our communities know that we can make things better because that knowledge affects our perception of the conflict. We are more likely to think defensively when we see a conflict as an unsolvable problem. However, when an issue is collectively viewed by a community as a moral injustice that can and should be resolved, then action is almost inevitable. The project of restoration rests on the assumption that people are generally motivated to act morally but face moral roadblocks related to moral development, bias, feelings of hopelessness, defensive thinking, and inability to perceive a moral conflict. What educators must facilitate is a shift from emphasis on the deficits in communities to their assets and potentials, from shame and blame to solidarity, and from determinism to hope.

As the discourse shifts, those whose needs are not met are not depicted as lesser or undeserving people; they are recognized as the targets of injustice. Stigma is replaced with moral indignation. Students should be outraged that anyone is being cheated out of health and education. They should be guided to positive solutions. The classroom must become a community development organization. All that stands in our way is our imagination and our perception of what is possible. Part of what is so wonderful about young people is that they have strong imaginations and haven't learned yet to believe that they can't make a difference. Youths are well positioned to be the leaders and free thinkers who will solve problems in their communities (e.g., hunger) where grown-ups have failed. I genuinely believe that a few committed educators and inspired elementary school students, with the support of their community, can end hunger in their neighborhoods.

Depending on the age of the students, shifting the discourse may not require a shift at all, but it may need instead to armor students against the change from a moral emphasis on care to one based on merit. As Hoffman (2000) discussed, such a change in emphasis happens at about the fifth grade. Children are often raised under value sharing, equality, and self-improvement until upper elementary grades, when considerations of merit, competition, and social comparison take over. This is marked by a broader shift in worldview, wherein children are more likely to view economic poverty as someone's fault than naively (and nobly) questioning how the world could allow people to go hungry. One might suggest that it is an important process for children to become adjusted to the "real world," from which parents often shelter young children. However, our concept of the real world, in this case, is socially inflicted and perpetuated by accepting the norm of social comparison and competition and ultimately our own degradation. We need to support the values that children naturally develop and prepare them to face the "reality" with the courage not to submit to a voice of morality hijacked by economic interests that serve the market rather than the people and planet. Because schools play a primary role in the shift in emphasis of moral principles, educators are well positioned to arm and armor students against the socially inflicted shift in moral emphasis and equip them with critical perspectives to encourage autonomous deliberation and choice rather than indoctrination. Additionally, engaging students in real-world problem solving would provide them experience to reflexively adapt and internalize their own principles.

For upper elementary and high school students, the shift in discourse would require conversation and experience. Students will be more motivated by affect and positive solutions than explicit knowledge. However, shifting the discourse can use explicit knowledge to launch conversation and facilitate experience. Just as the emphasis on social comparison begins to become prominent in the late elementary years, students are nearing adolescence, certainly a time of social comparison, but also a time when youths begin to test and develop their own moral principles, ethics, and political stances, often very idealistically. Such youthful idealism and adolescent rebellion are powerful forces and many of the most passionate environmental and social justice advocates whom I have met have been high school students. Thus, as problematic as I consider the socially inflicted shift from an ethic of care to principles of justice based on merit and social comparison to be, it's something that many youths should be well prepared to interrogate by late middle school or high school.

Discussion

> We…seek forms of knowing, teaching, and learning that offer more nourishment than the thin soup served up when data and logic are the only ingredients. In our complex and demanding worlds—inner *and* outer worlds—the human species cannot survive, let alone thrive, on a diet like that. (Palmer & Zajonc, 2010, p. 21)

The facilitation of discussion, as an academic exercise aimed at shifting the discourse, must attempt to broaden experience and understanding. The topic of discussion is no more significant than the way the topic is discussed. For example, lecture is a common method of discussion in classrooms. In most lectures, the teacher stands at the front and claims most of the airtime, occasionally encouraging students to participate but having them remain largely passive. Certainly lecture plays an important role in education, and it is not inherently a passive or pacifying experience—a good lecture can be "a well-staged theatrical drama" (Palmer & Zajonc, 2010, p. 29). However, a professor's "rigorous solo act…can feel more like *rigor mortis* from where the student sits" (Palmer & Zajonc, 2010, p. 29). I explore the value of a more open and active conversation among students. The teacher's role is to facilitate a conversation that is safe and academically productive and to provide clarification and direction to the conversation. In such conversations, the balance of power between teacher and student is shifted. Thus, both the content of the discussion and the way in which it is facilitated are beneficial learning experiences. The content of the discussion involves viewing a topic from multiple perspectives and expanding students' understanding of an issue, and the way the conversation is facilitated empowers students to make their own meanings and connections.

It is not permissible for the teacher not to actively engage with the students in the conversation or to rely entirely on student-directed conversation. The teacher should maintain authority without becoming authoritarian. All active participants should be recognized as co-teachers and co-learners. The teacher should be well positioned to help students build on their existing knowledge and connect their lived experience, their affect, and their tacit understanding, with explicit knowledge and curricular goals. It is necessary to incorporate a scaffolding approach wherein conversation is a central element allowing students to reflect together. This must be accompanied with curriculum emphasizing individual reflection and lessons or experiences for the students to reflect upon. If, for example, hunger is being discussed, students should initially be encouraged to voice their own knowledge, feelings, and experiences associated with hunger, but the educator must scaffold upon this knowledge of lived experience with experiences in the community, films, readings, and so on to guide students to deeper understandings.

It is important to emphasize the difference between conversation and debate. Debate is intentionally persuasive and competitive and explicitly a controlled form of conflict. Learning to form and articulate persuasive arguments and face social conflict in a formal setting is an important skill. However, facilitating debates is not a task that I have expertise or skill in, and in my own experience I've always feared that conflict and competition overshadow learning objectives related to the content informing the debate. Especially regarding moral conflicts, debate is unlikely to change anyone's original position. When someone argues their moral position and feel

> strongly about an issue, their feelings come first, and their reasons are invented on the fly, to throw at each other. When you refute a person's argument, does she[3] [*sic*] generally change her [*sic*] mind and agree with you? Of course not, because the argument you defeated was not the cause of her [*sic*] position; it was made up after the judgment was already made. (Haidt, 2006, p. 21)

There are certainly reasons for our moral judgments and Haidt (2006) is not suggesting otherwise in the above quote. Haidt (2006) is suggesting that we may not be able to identify or access those reasons and that our brain automatically attempts to make sense of our judgments. Conflict can become so emotionally charged that our arguments reduce to little more than confabulation (Haidt, 2006).

I don't want to suggest that debate, when not reduced to argument, is entirely ineffective. Debate is a particular skill and presents its own unique educational opportunity. However, I focus on the benefit of shifting the nature of the discussion from debate to open dialogue. In open dialogue, students should be encouraged to share thoughts, opinions, and feelings, but not in competition with one another or with the sole aim of advancing their opinion (or better yet, advancing the depth of their understanding by formulating sound arguments). The goal in open dialogue is to reflect and to learn from each other and to share an experience. Loeber, van Mierlo, Grin, and Leeuwis (2007) suggested that "an atmosphere of trust and commitment to reciprocity is essential" (p. 89) in such dialogue. The sentiment was "I'll let you in on my private considerations if you'll let me in on yours" (Loeber et al., 2007, p. 89). However, I favor a less strict demand for reciprocity (i.e., mature reciprocity) wherein everyone should have the opportunity to make a contribution as a valuable member of the group and be encouraged to do so by sharing their thoughts and opinions, but it should be OK not to speak, too.

In such a dialogue, the *why* (factual support) of a statement isn't always as important as the *what*. Asking "What makes you feel that way?" is significantly different than "Why is your statement true?" or even "Why do you feel that way?" Children and adults alike may not know, or may not be able to articulate clearly *why*, and they may either confabulate or respond that they do not know. As Loeber et al. (2007) suggested, "If there is no clear need to reflect on the tacit assumptions that underlie common patterns of behavior, they are often factored out of the discussion" (p. 88). However, if we ask *what* makes us/them say that, we may reduce confabulation and focus more on the conditions and emotions that lead to the judgment, which we can investigate further toward discovering a more explicit *why*. It is important to be clear, however, about the differences between *what* and *why* in this context.

People are entitled to their feelings and feelings are never wrong. However, our tacit assumptions and intuitions should be interrogated. By attempting to bring such assumptions to the surface by both questioning and encouraging questioning, we

> can reveal how social structures create stereotypes and lack of information that may lead to tension, alienation and conflict. Attentive teachers can invite those questions and affirm a classroom culture that creates trustful, respectful dialogue. (Bode, 2012, p. 344)

For example, in a discussion aimed at shifting the discourse on hunger, students may make certain tacit assumptions about the merit of those facing hunger. Through dialogue, "students who examine their stereotypes about why people need to make repeated visits to the emergency food pantry may begin to ask larger questions about why members of the community are consistently unable to provide for their families" (Eyler, 2002). We need to teach kids to value different ways of knowing, but to understand the different roles and function of those different ways of knowing. By exploring *what*, *why*, and *how* we know and feel through conversation and being clear about the distinctions, we facilitate growth, broaden our perspectives, and allow the group to delve deeper into issues.

Conflicts and disagreements may arise and can be resolved constructively as long as respect and equitable power are maintained. Through conversation, power is equalized. Berman (1997) suggested that the shift in power in open and democratic classrooms parallels civic life and encourages kids to become more involved in the democratic process:

> Educators should promote the belief that political and social change is necessary and good, that American society has problems (a "reality" students are well aware of); that there are means by which citizens can work to resolve those problems; and that when the channels of political change within the system are closed to certain groups, conflict (not necessarily violent) may emerge. Conflict should be treated as creative rather than necessarily destructive, because eventually it may open up awareness of

> participation to those who have been excluded. Conflict is thus a vital and constructive aspect of the American political system. (p. 115)

Although discussion is important, students need action to discuss and reflect on. Once a community need is identified and briefly introduced and explored, action is imperative. From that point forward, action and reflection must be regular and continuous. The discourse will shift as much with experience as with conversation. For example, a student questioning the merit of people continually needing emergency assistance "at the food pantry giving out emergency food boxes" eventually "arrived at the point where she is ready for critical reflection on fundamental principles about how society is organized to provide for its members" only after spending time volunteering at a local food pantry (Eyler, 2002).

Educational Action

> What students learn about poverty from texts is almost always less compelling than what they would have learned by doing the reading while volunteering in a community where the sights, sounds, and smells of poverty are inescapable elements of the educational experience. (Palmer & Zajonc, 2010, p. 31)

Shifting conversation about a community need into action is an educational process. I loosely follow the Action Research and Community Problem-Solving (ARCPS) model[4] (Wals, Beringer, & Stapp, 1990) to imagine what such a process may look like. The ARCPS model is fundamentally about empowering students as researchers and activists rather than as subjects of action research.

> Students are often unmotivated and dissatisfied in the classroom because they feel that what they are taught there is removed from, and not useful or valid in, the "real world." ARCPS attempts to address their concerns by challenging students with "real-world" problems. (Wals et al., 1990)

Additionally, students should benefit both internally (e.g., empowerment and motivation) and externally by addressing a community problem that affects them. Because ARCPS focuses on problem solving in the students' communities (or even schools), students may uniquely be subject, researcher, and beneficiary of action research. For example, action research conducted by students may focus on "misinformed decisions and policies intended to help them, but designed without their input." Youths can change this dynamic through action research and influence "the development of programs and policies designed to affect their lives" (Powers & Tiffany, 2006, p. S80).

The action research process begins with identifying a community need. The ARCPS model suggests that the issue should be of the students' choosing (Wals et al., 1990). This is certainly ideal and I think that teacher, students, and community must identify and define the issues together, but an educator can chum the waters to engage stakeholders. In such cases, the educator must be fully prepared to defer to emergent themes or concerns of students and community. An initial conversation should focus on the lived experience of the students (see the section "Discussion" in this chapter). Are they familiar with this issue? Do they think it is prevalent? Do they think it is fair? What feelings do they associate with it? This could be facilitated in a variety of ways and should be more informed by the teacher's relationship with the students than my hypothetical treatment. Centering student and community needs and emergent themes doesn't mean that the teacher doesn't come prepared for class; instead, the teacher must prepare for multiple possibilities and be ready to walk with the students in new directions or take authority and redirect students as necessary. All the while, the teacher must cunningly seek opportunities to make grade-level learning requirements relevant and applicable to the goals of the action research.

Once an issue has been identified, students should engage in research activities with emphasis on critical thinking and reflection. Fundamental to action research is reliance and emphasis on the involvement of community experts who "*live* the research issue" (Powers & Tiffany, 2006). Thus, students need to conduct their research in the community both acting as local experts and consulting community experts. Such research can and should be paired with readings, films, and the like in the classroom.

> By promoting critical thinking and the exploration of the social circumstances related to research questions, participatory research goes beyond mere fact-gathering and report-writing and uses the knowledge gained to guide and energize collective change in communities, organizations, programs, and the research participants themselves. (Powers & Tiffany, 2006, p. S79)

The focus should rapidly shift from an emphasis on understanding the problem to focusing on possible solutions and opportunities to take action. It is important that initial action take place as soon as possible, even before full commitment or deep understanding.

> This approach differs from the standard way of solving a problem, which is to first thoroughly understand the problem then to consider what actions, if any, to take. Evaluation is normally carried out only after a plan is implemented. The drawback of this method is that it tends to delay action while endless study is done, and it fails to recognize that one often comes to really understand a situation only through taking action and evaluating the consequences of that action. (Wals et al., 1990)

Action cements an issue in the here-and-now, which we have empathic bias toward. However, before becoming too involved in initiating a community project, it is important to get to know the community and the issue, especially the current stakeholders and what is already being done. The process can be summarized as plan, implement, evaluate, repeat, but rather than that particular order, beyond an initial cycle action and reflection must be simultaneous and constant. Students need to reflect and evaluate individually and collectively, and should have the opportunity to act individually as well as collectively.

Finally, as Anguiano et al. (2012) pointed out, there is no cookbook approach. The nature of the methodology is that it grows and adapts itself to the participants and to the issue being addressed. Action research and community problem solving are more about an approach to the process than specific methods. This is an approach to education that centers on community and empowers students and community members by building relationships and balancing power among stakeholders in the process of understanding and solving problems. It is also about building relationships in order to address a specific community need. "It is intended to be a social process that is participatory, practical, collaborative, emancipatory, critical, and reflective" (Anguiano et al., 2012). Although there are some subtleties in the specifics, action research, critical pedagogy, popular education, integrative education, and multicultural education are among a number of other named approaches aiming at a more holistic, empowering, and consciousness-raising education that demands action.

Application of Pedagogy of Basic Needs

> Although intrinsically an important and worthy topic by itself, school food can also be an entrée (so to speak) into many other topics of interest to education researchers. It can be the highlight or simply a small piece of context. It can be a central focus or simply one among many facets to explore as a possible confounding influence.

> Whatever the relative weight of food in a study, it is time to take school food seriously, to consider how much depends on this most human, embodied part of the school day. (Weaver-Hightower, 2011)

I've very generally outlined a theoretical approach to problem-solving education, with specific consideration of hunger, and emphasized that by changing the way we discuss problems, we see solutions differently. Although the actions and solutions should be co-constructed with students, I provide a vision of what implementation of the process might look like with specific examples from real community projects that address community needs, support moral growth, and meet learning objectives. The premise for this example is that several kids in a class (or other setting such as an after-school program) experience food insecurity and students have come to realize that hunger is prevalent in their community and want to do something.

A case against charity

Before continuing to explore the wonderful things that folks are doing to address food insecurity in their communities, I want to consider a common way that it is addressed in schools: canned food drives. Food drives do nothing to destigmatize hunger, and often imply that hunger is an external (in the community) while ignoring or overlooking internal (in the classroom) need. When I was in elementary school, the class that donated the most cans in our school food drive got to have a party. For several weeks, a tally on the board kept track of how many cans had been donated and everyone knew who the big can donors were. The number of cans a student had donated became a measure for social comparison. Although having enough to eat was never a concern in my family and my parents sent me to school with several cans of food, I distinctly remember sneaking additional cans out of the pantry to bring to school. To be clear, I didn't sneak more cans because I was engaged in contributing to a humanitarian effort; I snuck them because I wanted to have a party and because I wanted to be recognized as a contributor.

I wonder how kids who couldn't bring any cans felt. I wonder how many kids donated food knowing that it would mean they wouldn't get enough to eat in order to avoid being shamed because their family was unable to donate. This is a process that adds shame to poverty when the goal should be exactly the opposite. It is demeaning to students facing food insecurity to conceptualize food drives as charity rather than justice. We tend to celebrate charity, as if social change is something the privileged do for the "underprivileged" who are unable to help themselves or each other. This type of charity is little more than self-congratulatory social posturing for privileged elites and it doesn't have any place in action research. However, neither canned food drives, nor the often genuinely humanitarian goals of charity are innately harmful. It is simply a matter of how we approach helping one another.

We shouldn't do charity *for other* communities. This paternalistic way of helping is often associated with charity that encourages downward social comparison and stigmatizes need, but it is not the only way to conceptualize charity. We need to work with/in our communities[5] in solidarity for justice. The point I am trying to make is that good intentions alone are not enough. We have to focus on *how* and *why* we help just as much as *what* we intend to do and we must always work *with* rather than *for* those we help. Thus, the approach requires humility, respect, and attention always focused on the outcomes of the work of both participants[6] and beneficiaries (who should be one and the same).

A case for educational action

What if children come to see food insecurity as injustice? Need would begin to be destigmatized and no one would be made to feel ashamed of being hungry. Students could work together to make sure that they and their community had enough to eat, not as charity, but in a spirit of solidarity and community. Students could work together to do something as simple as implement a "share bowl" in which students could leave extra food from their lunches. Anyone would be free to leave or take something whenever they wanted. It's absolutely atrocious to see how much is wasted in a single lunch service in most school cafeterias. This is something very simple that I have seen implemented in school cafeterias to reduce waste and hunger in schools. Other schools and communities have implemented their own pantry programs and have started sending kids home on weekends with a backpack full of food (cf. Rodgers & Milewska, 2007). The internal need in schools must be addressed before attention is shifted to external needs.

Once the immediacy of the problem is sufficiently addressed and the class is able to consider the issue more structurally, students could begin planning, implementing, and evaluating solutions in and with their community. One solution might be to implement or expand the school garden for the purposes of donating to the food bank, supplementing school nutrition, establishing a school-based food bank, or even preparing and serving free meals. They could potentially partner with churches, community service organizations, or police and fire departments. One thing that is great about kids is that they are adorable and communities are often incredibly inspired by the efforts of their young people and eager to support them.

Action research toward food justice as part of a curriculum with a food education theme could easily meet learning goals while providing a meaningful experience and solving a real problem. An educational project with the goal of eradicating hunger in the community could also inspire a sense of meaning and purpose, build community, and encourage civic engagement and moral growth reaching far beyond the classroom. Students could learn about healthy diets and lifestyles, food systems, and botany by learning to cook and garden. These skills could be applied to their project addressing community hunger and they, in addition to expanding school gardens, could work to establish community gardens, plant gardens at home, and in the process begin establishing decentralized food systems in their communities. They could donate excess produce to a community pantry or open their own. They could also prepare and serve meals in their school or community. In the course of such efforts, students could explore the diets and agricultural systems of other cultures and their own in social studies classes. In science classes, they could study the biology of plants, the water cycle, photosynthesis, and nutrition—all knowledge relevant to their project. Through spending time in the garden, students could learn natural history and ethnobotany and more generally connect with nature. Math classes could directly apply to nutrition, cooking, and gardening.

Not only a classroom but also an entire school could engage in a student-led action research project. Every class and every subject could be related to the single goal of eradicating hunger, and students at all grade levels could engage in the process. Not only would it be possible to meet all of the standards in this process, but the students would also be eager to learn skills that apply to solving an important problem.

Examples of action research

Students have accomplished amazing things through participatory action research. One action research project that focused on youth homelessness in upstate New York recruited youth researchers from a community program serving runaway youths (Powers & Tiffany, 2006). The youths were involved in the design of the project, data collection, interpretation, and the presentation of findings and recommendations. One difficulty with studying homelessness among youths is that they tend not to use support services as frequently as do adults, and researchers often have trouble estimating the population of youths experiencing homelessness, much less engaging with them (Powers & Tiffany, 2006). By involving youths familiar with this population through their own lived experience, the research team was able to conduct 165 interviews with youths experiencing homelessness. They presented their reports and recommendations to community stakeholders including legislators, funders, human services staff, university researchers, and policy makers. Thus, the youth researchers became powerful advocates for their peers experiencing homelessness in a way that the detached researcher likely could not.

Nygreen, Ah Kwon, and Sánchez (2006) described several fully youth-led projects that developed organically out of existing relationships between the authors and youths. Nygreen et al. (2006) considered themselves adult allies, outlining a strictly supportive role. One project involved current students and recent graduates of a public alternative high school predominantly serving low-income Students of Color who had been labeled "at risk." The group's purpose was to make changes in their school. For a year, the group held weekly meetings to discuss social issues and personal experiences and conducted inservice programs and interviews with students, teachers, and staff in an attempt to understand how social issues affected their lives and educations. The following year, the students and their "adult ally" developed and instructed weekly eighty-minute lessons in the school's U.S. Government class designed to encourage "students to think critically about social justice issues and engage in action for social change" using "dialogue-based pedagogy to address topics like police brutality, prisons, and environmental racism" (Nygreen et al., 2006, p. 111).

Another group of four Latina transnational second-generation immigrants including one graduate student and three high school students who all regularly spent time with family in Mexico spent a year sharing their pictures and experiences (Nygreen et al., 2006). A frequent discussion topic was how their experience, language, and culture were ignored in school. This inspired the group to write a children's book on what life is like having homes on either side of the border. Beyond their own experience, the group conducted research through dialogue, interviews, a collection of pictures, home videos, and other documents from transnational families. Using this information, they wrote and illustrated their book. They initially printed eighty copies through a local small grant and distributed the book to their families, at conferences, local libraries, and day care centers. At the time of publication of the description of the project (2006), the group was working with a children's book publisher for national distribution.

I hope to have illustrated through these examples that education focusing on the most immediate community need can meet learning goals, solve problems, build community, and facilitate moral growth. This may seem too idealistic, but I argue that restoration must be idealistic. We can't replace idealism with practicality; instead, we have to remain idealistic in order to guide our practical efforts. If our ideal and our practice match, we have little direction for progress. Thus, the central theme of this section is rethinking education to start from the bottom of Maslow's hierarchy and to work our way up. I've argued for a radical approach; our duty as moral humans and as educators is to do what we can in the spaces where we work. Different institutions and different organizations have different capacities. It's often slow and difficult work, but if we're not in this to serve kids, then we're not in it at all. I realize this is easy to write and difficult to do, but I want to conclude by sharing some my own recent experiences observing lunches in school cafeterias before and after salad bars were installed. The principal dressed in a banana costume and danced around the cafeteria to encourage kids to eat healthy foods. The food service manager quietly paid out of her own pocket for lunches for kids who were behind on payment.[7] Food that had been grown in the school garden was served in the cafeteria. I saw everyone from parents, to teachers, to administrators wanting to do better for kids and working to accomplish it. I saw kids genuinely excited to eat their vegetables! I saw hope. The point is that we need to be doing big things and making radical changes, but all of the small things we do along the way matter.

Extending to Ecological Issues

Action research applies beyond immediate physiological and safety needs. It is a model for education as social change. This could mean establishing more green space, reducing litter, recycling programs, and the like. The sentiment of a pedagogy of basic needs is that the most immediate physiological and safety needs must come first, or occur simultaneously with addressing less immediate issues. We must help all people meet their basic needs in order for them to rightfully claim their place at the table in discussing the large-scale structural changes necessary in the face of issues such as climate change.

Ecological degradation is readily identifiable as a significant impairment to survival needs in some parts of the world. As such, degradation continues (e.g., climate change), the conflict between humans and nature will escalate, and social reform, as it has in many parts of the world, will not be a utopian dream, but a matter of survival. Conflicts between meeting our basic needs and preserving the environment are created through economic exploitation that reduces all of life (human and planet) to the status of resource. This is killing us. In a struggle that frames human versus nature and human versus human, we divide and conquer ourselves in the face of our adversaries (nature and each other). This may be the harsh verdict of nature that Bookchin (1982) described. When we come to see the struggle as human with nature and human with human, then there is no adversary at all. We need to meet the immediate needs of people and planet, but we can simultaneously circumvent the processes that are leading to folks having their needs unfulfilled.

Meeting Basic Needs Mandates Social Change

Social change can be a survival need when social structures impede one's ability to meet one's physiological and safety needs. I imagine such social structure as a competition to see who can tread water the longest, when some people have water wings (privilege) and some have an anchor chained around their waist (oppression). It could be suggested that everyone has a fair chance to win based on skill and endurance (equal opportunity). This might be easily accepted by those with water wings, and reluctantly accepted by those with anchors (internalized oppression). But what we don't realize is that we'll all become exhausted and drown eventually if we keep playing zero-sum games in which our success is inevitably at someone else's expense. We all need to help each other out of the anchors and share the water wings (meeting basic needs) with the goal of getting everyone out of the water (radical social change).

As Maslow (1943) wrote:

> There are certain conditions which are immediate prerequisites for the basic need satisfactions.… Such conditions as freedom to speak, freedom to do what one wishes so long as no harm is done to others,…justice, fairness, honesty, orderliness in the group are examples of such preconditions for basic need satisfactions. Thwarting in these freedoms will be reacted to with a threat or emergency response.… These conditions are defended because without them the basic satisfactions are quite impossible, or at least, very severely endangered.

The Zapatista uprising exemplifies the necessity of social change to satisfy basic needs. The Zapatistas identified economic policy (NAFTA) and the dissolution of communally held land as not only destructive to their way of life, but also as an attack on their means for survival. For the Zapatistas, NAFTA was perceived to be a death sentence. The Zapatistas realized that meeting their basic physiological and safety needs was impeded by larger structural issues.

The Zapatistas have consistently worked for structural change and inspired an international Zapatismo movement with constant emphasis on the well-being of the people over armed revolutionary conflict. The Zapatistas have worked to autonomously meet the needs of their own communities by communally operating farms, schools, and clinics (Hayden, 2002). Many other revolutionary struggles have failed to place the highest value on life and have considered that the ends justified the means. When this happens, even a win is a loss and one brutal dictatorship replaces another. The only revolutionary struggle is one of and for those who love and honor life more than power.

Pedagogy for restoration is divided into three components strictly for conceptual clarity. The ultimate goal is planetary restoration. Pedagogy of identity and pedagogy of purpose circle back to the problem-solving process and constant attention to meeting everyone's basic needs and the necessity of large-scale social change. Certainly groups and individuals whose basic needs are better met can engage further with meeting the basic needs of larger human and non-human communities and work toward ecosystem-scale restoration. Through pedagogy of basic needs, identity and purpose already begin to be addressed and pedagogy of identity seeks to illuminate purpose. In turn, pedagogy of purpose refers to pedagogy of basic needs, because when one finds purpose and self-actualization, they have a duty to act.

· 7 ·

PEDAGOGY OF IDENTITY

> Native students confined to learn only in Eurocentric systems have poor success rates. The experience can be described as "looking into a mirror and having the mirror look away." (von Thater-Braan & Nelson, 2012)

Once we've fed our students, we need to know them (to the extent that it is possible to know someone else and their experience) and help them to know themselves and each other. Because to know oneself is to know how one fits into the world, and to become critically conscious of the world. After this, attainment far beyond academic and economic success is inevitable and the practice of teaching is inconsequential because students will be mentally and physically prepared to learn and become open to education in all facets of life.

The epigraph to this chapter refers to Native students' experience in contemporary public education. Before continuing with this discussion of the importance of honoring identity in education, I want to share the narrative of Sun Elk from Taos Pueblo who attended Carlisle Indian Industrial School for seven years beginning in 1883:

> They told us that Indian ways were bad. They said we must get civilized. I remember that word too. It means "to be like the white man." I am willing to be like the white man, but I did not believe Indians' ways were wrong. But they kept teaching us

> for seven years. And the books told how bad the Indians had been to the white men—burning their towns and killing their women and children. But I had seen white men do that to Indians. We all wore white man's clothes and ate white man's food and went to white man's churches and spoke white man's talk. And so after a while we also began to say Indians were bad. We laughed at our own people and their blankets and cooking pots and sacred societies and dances. (as cited in Dunbar-Ortiz, 2014, p. 212)

Sun Elk and countless others eventually returned to their native way of life, but reintegrated with much difficulty (Dunbar-Ortiz, 2014). This devaluation of non-dominant culture and coerced assimilation covertly continues in schools today and is a source of internalized oppression.

This chapter focuses on Maslow's psychological needs of love, belonging, and esteem. It encompasses helping students explore their own identity (broadly, cultural and ethnic identity, but also specific identities, e.g., environmentalist). It is also about ensuring that students find themselves in the curriculum and that it speaks to their own experience (e.g., culturally relevant history). Pedagogy of identity is not only a process of self-discovery, but also a group exploration. Students should learn about one another's identities and experiences and build relationships recognizing that while they are each uniquely different, they are similar in many important ways. Most important, they should learn that we are all connected and come to respect and appreciate U.S. American cultures and world cultures. It is critically important that educators engage in the process, letting students know them as they work to know their students.

I broadly consider identity to be the culmination of experiences, statuses, traits, and beliefs that make up who people are, including both our own self-conceptions and the way others perceive us. "Identity is linked with self-concept and self-reflection, and includes beliefs about who we are as we are as well as who we hope to be. It is, in part, a way of locating oneself in the world" (Clayton & Opotow, 2003, p. 299). Thus, identity must be "understood as a complex and multidimensional construct with motivational and behavioral implications" (Clayton & Opotow, 2003, p. 300). Individuals are composed of multiple identities, sometimes conflicting and sometimes confusing. Some elements of identity are biologically determined and some socially determined, some claimed by ourselves and some imposed by others (i.e., attained and ascribed identity). However, our identity is more than the sum of its parts and while culturally constructed markers of identity such as gender, race, and class may impose meaning about who one is (i.e., ascribed), they are only one dimension of identity and do not define anyone entirely[1] (Clayton & Opotow, 2003).

Identity is often comparative: by claiming an identity, or having one imposed, an implication is made that this label significantly differentiates the person claiming the identity (or having it imposed) from others. The implication is that if someone significantly *is* something, then someone else significantly *isn't*. This establishes an affinity among the group of individuals to whom the label applies and a distance from those to whom it does not. It is a natural human tendency to attempt to understand, categorize, and even stereotype. This behavior is not inherently harmful. However, the politics of identity are quite complicated and value laden and identity can be used to establish hierarchy, to divide, and to oppress (see the section "Social Death" in Chapter 5). It can also be dismissed entirely in a way that effectively erases people, their experiences, and voices (e.g., colorblindness and forced assimilation).

It is necessary for all educators to help folks celebrate their similarities and their differences. Taking the importance of similarity or difference to the extreme is almost always detrimental. However, the relative importance of similarities and differences depends on context. In certain contexts, it may be important to focus more on the ways in which we are similar and to stand together in solidarity (but never in a way that attempts to completely homogenize). In other cases, it may be important to focus more on the ways that we are each unique individuals with different experiences. For example, in the struggle against oppression, we must center the identities and statuses most frequently targeted by oppression and determine how and why they are targeted.

As we explore identity (especially cultural identity), we must carefully avoid overgeneralization. One's identity reflects some shared experience among that particular identity but common experience does not dictate shared values or beliefs. However, cultural identity may influence dominant themes in beliefs and values (Moya, 2003). Epistemology can be culturally constructed; thus, not only the experience but also the theory through which we interpret experience is influenced by cultural identity (Moya, 2003). The task is to avoid making assumptions about any individual's values or beliefs based on cultural identity, and to recognize that cultural identity can influence values and beliefs. Our self-conception is relative to culture but it is not entirely culturally defined. This is why, in education, the students' own exploration of identity and the relationship between teachers and students is so important. We need to know (to the extent that we can know, but never become entirely familiar with or entitled to) our students' unique identities and cultural experiences and the intersection between their own cultural identity and dominant culture. Furthermore, neither identity nor experience is static and their associated meanings are subject to change at individual and collective levels.

Role Models

Chuck D discussed the important role of popular culture (Chuck D & Jah, 1997), especially of Black[2] musicians, in fighting racism. Chuck D argued that all Black public figures have a duty to strive to be positive role models.

> The athletics and entertainment industries have both had a tremendous impact on the mind-set of today's youth to the point where many young people grow up idolizing athletes and entertainers, yet the only time these "idols" are seen is if they're playing ball, singing, rapping or dancing.... Black youth, especially young Black men, gravitated to Rap when it first came out, because it's a vocal expression, whereas in athletics you'd see somebody do incredible athletic things on the court or field, but you'd never hear anything from their mind. The travesty is that both athletes and entertainers could be and should be, more visual and vocal sources of hope for young children in the ghetto and throughout the world. (Chuck D & Jah, 1997, pp. 95–96)

Chuck D further argued that Black people, especially Black public figures, model behavior for Black youths and broader public perceptions and stereotypes (Chuck D & Jah, 1997).[3] Thus, he suggested that when public figures glamorize violence and material wealth, they normalize such goals for youths and furthermore contribute to racist stereotypes. Regardless of whether this is right or wrong, it is real.

> It's more than just saying, "I'm not a criminal or contributing to a criminal element that should be good enough." It's not enough. It's not an issue of donating money to Black causes. What makes me pissed off is that they [Black athletes] are not saying much of anything. Their words and their time could steer a lot of young people out of the traps that are set for them. (Chuck D & Jah, 1997, p. 100)

All public figures have an obligation to pay careful attention to their public perception and persona.[4]

No one should be so presumptuous as to believe that their behavior reflects on only themselves. This presumption is a significant function of privilege because whites are rarely burdened with an awareness that our actions could be taken as a reflection of whites with whom we don't consider ourselves associated. However, we are all co-responsible for bringing about social change. We must all be aware of both how we see ourselves and how others see us. We are all role models and ambassadors (at least sometimes) for the identities we are socially assigned and the collective identities in which we claim membership.

People of Color alone should not be burdened with the task of breaking down racial stereotypes and the work of anti-racism. I insist that we challenge the expectation that People of Color should be accountable for their entire (socially constructed) race and constantly expected to prove that they are "good" (see the section "Entitlement and Deservingness" in Chapter 2). However, as long as such an unjust expectation remains prominent, whites must hold themselves to the same (unjust) level of accountability. An example of this unjust disparity relates to the election of Barack Obama, our first African American president. Obama had to prove that he was a U.S. citizen (not a Kenyan), not a terrorist, and not Muslim—all three criminalized statuses rather than inherently criminal identities. Such scrutiny has never been so publicly prominent for any other president of the United States. Someone with the status of U.S. Senator and then President of the United States of America faced pressure to establish value not only as a politician, but also as a human being. The underlying injustice is that privileged identities and statuses are presumed to be good and entitled to rights while dismissed, marginalized, and criminalized peoples are required to prove that they are good and deserving of rights (Cacho, 2012).

As much as I do not like to claim identities such as white, man, and environmentalist or a status such as citizen, because of the privilege (and certain behaviors) associated with those identities and statuses,[5] I recognize that I have significant power in challenging the associated meanings. Rather than renounce my own privileged identities, I claim them and challenge their definitions by the simple fact that I refuse (or aspire to refuse) to resign to privilege and stereotypes and I interrogate the ways in which I am conditioned by those identities. It would only be from a position of significant privilege that anyone could claim not to be accountable to a socially imposed identity (e.g., white or Black). As long as such an unjust disparity exists, I will (aspire to) hold myself similarly accountable. Furthermore, when it comes to more radical stances relating to race and racism, as a white man, and a unique individual, I say things differently and I am heard differently. Equality is a utopian idea, but on the path to utopia we need to focus on equity. Given that we do not have equal agency and privilege, we need to think about what is fair and how we can each contribute to a better world uniquely, individually, and collectively.

Oppression is violent and acts of oppression may be traumatic to both those who enact oppression and the targets of oppression. The colonization of America was and continues to be traumatic for colonized and colonizer alike and it's not too late for us to collectively experience posttraumatic growth. Those at each level of collective and individual identity and in the intersections of multiple identities have a duty to work in any way they can to reduce violence in the service of meeting the ultimate goal of ecological and social restoration. I don't believe that People of Color and whites exist as a binary framework of race and relation (us or them) where whites committed to being anti-racist can be considered allies at best and saviors[6] at worst. This also applies to other dichotomously situated identities and statuses related to sex, gender, sexual orientation, citizenship status, and educational attainment. I believe there is a time and place for allies and saviors; however, I want to stand in solidarity as a co-conspirator in an insurgency against all oppression and a struggle to bring another reality to fruition. I want to be clear that I am not suggesting that as a white male I feel entitled to be present, influential, or even welcome in every space. Furthermore, we all need to hold each other and ourselves accountable while simultaneously practicing acceptance. What we must not do is reduce all identities and experiences in life to a single human experience.

Hispanic Environmental Identities

Anguiano, Milstein, De Larkin, Chen, and Sandoval (2012) studied the intersection of environmental values and beliefs regarding the natural environment with those of dominant culture in New Mexico. The dominant ideology of environmentalism is so prevalent that Anguiano et al. (2012) found a "deep disconnect between the word 'environmentalist' and Hispanic communities unwilling to ascribe to such a label despite practicing sustainable ethics" because

> Hispanic community member participants may already practice social and environmental sustainability, but…their strategies are tied to culturally specific values and meanings and thus differ from those advanced and advocated by the dominant

> environmental movement...despite the fact that their articulated behavior was largely analogous with the environmental justice movement's aims. (Anguiano et al., 2012)

Anguiano et al. (2012) significantly raised the point that it is not fair to assume that ecological or social sustainability is not a priority for those whose practice may not align with the dominant environmental movement.

Educators and mainstream environmentalists must acknowledge different cultural conceptions of the environment.

> Anglo-American environmental discourses that assume nature is an untouched space, that "wilderness is pristine," and that "protecting nature means protecting it from predation by humans" do not resonate with Latino communities. (Anguiano et al., 2012)

Many of the Hispanic community member participants in Anguiano et al.'s (2012) study lived and worked closely with the land, including many ranchers with long-standing generational ties to the land.

> In contrast with the individualistic and culture-nature binary view of dominant Western environmental discourse, for our Hispanic New Mexican participants, their environmental concerns stemmed from close ties to their community, neighborhoods, and place. (Anguiano et al., 2012)

This sense of place is quite complex, considering that the

> deep history of colonization[7] implicates New Mexican Hispanics as both colonizers and colonized, yielding a complex mixed race identity (e.g., indo-Hispanic, mestizos, and anglo-Hispanic) that results in interesting paradoxes and views about land ownership. These paradoxes partly can be attributed to more recent patterns of exploitation exercised since the 1848 Treaty of Guadalupe Hidalgo ended the U.S.-Mexican War and ceded almost one half of Mexico to the US, with the Southwest Hispanic colonizers essentially repositioned as the colonized.

New Mexico is also home to many Latinas/os who (or whose families) immigrated after 1948 and the influences of both Spanish and U.S. American colonization are prevalent throughout the state. Additionally, since 1948 the government has seized community-held land from Latina/o communities and dispossessed Latino/as of lands along the U.S.-Mexico border (Chen, Milstein, Anguiano, Sandoval, & Knudsen, 2012). One begins to see that broad generalizations about the values and beliefs of Hispanic peoples even in the specific context of New Mexico are impractical. However, attention to the collection of experiences of these peoples is important in understanding the formations of individual values and beliefs.

An example of how "environmental justice"[8] organizing may look for Hispanic peoples in New Mexico illustrates the intersections of individual and collective identities and hints at the construction of new collective identities. A volunteer with a group organized to prevent and eliminate environmental and community degradation by a local sawmill recalls becoming involved with the group:

> L: So within like 15 minutes I was the vice president and I was like, what is going on here? "And by the way, we are going to have enchiladas you need to go home and get all this food." and I was like, whoa!...One thing we understood is that every time we had an event, there was some kind of celebration; we had music; we had food. We understand the way our people think: food and music. There was a lot of posole[9] that we made.
>
> This participant attributed community activist success to understanding that involving people meant having posole, music and community engagement so that the fight against environmental injustices had a comfortable and familiar feeling. The notion of understanding "our people" speaks to an activism that recognizes, caters to, and grounds neighborhood organizing in its cultural traditions rather than imposing a foreign framework that would not resonate with the participants. (Anguiano et al., 2012)

Thus, Anguiano et al. (2012) discussed the importance of not only literally speaking the language (e.g., Spanish) but also speaking the language of firsthand knowledge of culture and tradition. "The larger environmental movement must honor the voices of local communities and acknowledge the importance of linking cultural orientations within the larger environmental agenda" (Anguiano et al., 2012). Although as educators we can never be completely familiar with any culture but our own, we must be genuinely interested in and respectful of the people we work with and in their language, culture, traditions, and experience, and in working together in solidarity, for a better world.

Additionally, new collective identities can be joined, expanded, or formed without threatening our existing individual identities. An exemplary case of collective identity is the Zapatistas. The iconic image of the Zapatista with black ski mask, Cuban military cap, and red bandana is widely recognized around the world. By hiding their personal identities, they portray themselves as one. As a mestizo, Subcomondante Insurgente Marcos's lived experience is not that of the Indigenous Mayans. As the story goes, Marcos traveled the Chiapas, encouraging and organizing an uprising informed by Western political theory but he was initially unsuccessful because he lacked a deep understanding of Indigenous culture (Hayden, 2002). The rumors concerning his identity, education, and background are rampant, but one thing is for sure—when he puts on the mask and adornments of the EZLN, he is absorbed into the collective identity of the Zapatistas and is neither Indigenous nor mestizo. Furthermore, Marcos has proclaimed that anyone who stands against injustice is a Zapatista.[10] In this way, the collective identity is not exclusive to Mayan people. While the EZLN retains primarily Mayan leadership and attention to Mayan interests in Mexico, it is done in recognition that their local struggle is part of a worldwide struggle against oppression.

Thus, the Zapatistas balance the need for a single collective identity and voice individual identities. One can dedicate oneself to a collective identity associated with a social cause without sacrificing the individuality. Participants maintain their individual identities and gain a collective sense of identity. We need that sense of collective purpose, identity, belonging, and acceptance, and we need our sense of individuality. We need to reinforce both, through education, and emphasize how connected we all are in the collective identities of humanity and of life on Earth.

Hip Hop

> They tell lies in the books
> That you're readin'
> It's knowledge of yourself
> That you're needin'
> —"Prophets of Rage" (Chuck D & Jah, 1997)

Hip Hop is a prominent American (but not exclusive to America) culture and thus Hip Hop must be part of American education. The products of Hip Hop (e.g., albums) foster a certain experience in those who consume them, but the culture of Hip Hop is lived experience. Folks who do not share the lived experience (such as myself) cannot claim to know it through consuming the products. However, we can and should interpret the products empathically. The music and the visual and performing arts of Hip Hop (and the arts of many cultures) are deeply subversive, defiant, and educational while simultaneously being emotional, poetic, and beautiful. This approach to Hip Hop (and Hip Hop itself) can be applied to the exploration of other cultures and facets of identity. Thus, the exploration of Hip Hop is intended to illustrate an application of pedagogy of identity.

Rather than suggesting that educators bring Hip Hop and non-dominant popular and traditional cultures into the classroom to engage students and meet learning goals, I suggest that culture (i.e., Hip Hop) is already in the classroom and doesn't need to be assimilated into the critical educator's pedagogy. To say that Hip Hop is critical pedagogy or to say that Hip Hop is postmodern (cf. Potter, 1995) or for that matter to make such statements about any culture is to suggest that it be valued in its relation to the value of dominant academic theories.

> Black history carries the subversive truth that contemporary rationales for poverty, ghettoization, and trickle-down economic policy that justify the increasing wealth of a few on the backs of a growing black underclass are part and parcel of this ongoing capitalistic hegemony, and black arts are the signal site for the return of these repressed realities. (Potter, 1995, p. 7)

Thus, Hip Hop is already its own critical pedagogy, discourse, and epistemology.

> Coming of age outside of the schoolhouse and as an expression of lived curriculum, hip hop resists the images of American national identity presented as official school curricula. As lived curriculum, hip hop signifies a site of cultural struggle where the politics of race, class, sexuality, gender, place, and generation (among other subjectivities) are played out through the discourse of power, privilege, and oppression. (Baszile, 2009)

Hip Hop stands on its own as a culture grown within the dominant Western culture and from its inception has been critically conscious and anti-oppressive. I intend to illustrate the unique and independent value of Hip Hop and suggest a synergistic relationship with critical pedagogy.

Certainly the concern could be raised that the language and content of some of the most influential works of Hip Hop are not appropriate for the classroom.

> There are, no doubt, expressions of sexist, racist, homophobic, capitalist constructions; however, there is also a disruption of these oppressive discourses, often within the same performances. (Baszile, 2009)

I would passionately argue that we need to welcome that conflict in the classroom and address it because by censoring Hip Hop products we marginalize those whom Hip Hop speaks to and whom Hip Hop represents. Hip Hop is deeply and necessarily contradictory, complex, nuanced, and perfect in its imperfection because it is reflective of the contradictions of the society in which it exists. It's not refined and sanitized like academic work and that is how it is supposed to be. For exactly those reasons, Hip Hop is reaching and moving folks who academics don't reach and won't move.

> Ultimately, the critical point here is not—I believe—whether hip hop is good or bad, but that it in essence represents the struggle for freedom (even when acted out in negative ways) on behalf of young people who negotiate and perform self-meaning within a largely different sociopolitical context from previous generations. (Baszile, 2009)

Shakespeare v. Tupac

Using Hip Hop education is not just about how teachers can use Hip Hop products to engage students. It's about what educators and students can learn together from Hip Hop culture. Hip Hop is knowledge and movement, facts and feeling, explicit and tacit (KRS-One, 2009). Education has the Hip (i.e., knowledge—but not always the important and culturally relevant knowledge) but it doesn't have the Hop (i.e., feeling, movement, action). It's not action oriented; it's banking oriented (depositing knowledge for economic success).

Obligatory readings of Shakespeare in high school English classes are not generally framed as a way to get students interested in reading, but as the study of important canons of English literature. However, such canons of a dominant and domineering culture don't speak to many youths but instead derogate them or enforce their invisibility. The discourse of literary canons seems to render traditional culture historical and static and popular culture as contemporary and fleeting. What remains as canons are the works that speak to idealized notions of timelessness. However, as Duncan-Andrade and Morell (2007) pointed out, popular culture has always been a topic of academic inquiry for the academic elite.

Maybe rather than making Shakespeare relevant by aligning the works with modern popular culture, we should be studying the canons of the cultures that students affiliate with, including traditional cultures and contemporary popular cultures. I say we replace (or at least augment) Shakespeare with

Tupac, if Tupac speaks to our students' own experience. To do anything else is to weaponize literary canons as tools of colonization by enforcing dominant culture and ignoring the important and relevant works of non-dominant cultures. As I have already argued, the knowledge of and from non-dominate cultures found in such works provides critique and counter-examples to the degradation attributable to dominant Western culture.

Regardless of idealistic notions of what education should be, educators have a duty to balance the important work of facilitating meaningful education with the instruction necessary to help students run the gauntlets of the academy such as standardized testing. Duncan-Andrade and Morell (2007) are teacher-researchers who use popular culture in their English classes. They wrote: "Regardless of our philosophical foundation, we understood that our students existed in a world where they would be expected to take and perform well on standardized tests that served as gatekeepers to post-secondary education and, by consequence, professional membership" (p. 185). I argue that education relevant to students' own experience focused on identity and elements of popular and traditional culture meets learning objectives just as pedagogy of basic needs does. More important, pedagogy of identity, as explored through the example of Hip Hop, helps students explore their own identity and become empowered by an education that does not render them invisible rather than marginalized by one that does.

Duncan-Andrade and Morell (2007) pointed out that in order for students to effectively critique power structures and become critically conscious, students must understand the language of power and familiarize themselves with the dominant texts. Thus, studying the literary canons can be a subversive effort and studying culturally relevant materials could provide tools to critique dominant works. However, such a critical reading of dominant texts does not prevent students from making connections between these texts and their own lives. In fact, part of a rich academic conversation could focus precisely on exploring if/how students relate to the literary canons.[11] Duncan-Andrade and Morell (2007) wrote that they were only able to motivate some students to engage with such literature through centralizing a commitment to freedom and social change.

Wouldn't some students inevitably feel left out?

Pedagogy of identity is about exploring identity together and certainly seeks a pluralistic approach to identity. This means questioning Eurocentric dominance in the classroom and working to more equitably represent non-dominant cultures. Only dominant, racist ideology leads to the suggestion that an exploration of Hip Hop would speak only to those most severely marginalized by dominant Western culture and exclude students of privileged identities. Everyone has culture, and everyone benefits from exploring their own identity, cultural experience, and ethnicity as well as the diverse identities present in the classroom and in broader communities. Building such relationships with oneself, one's culture, and broader communities connects and empowers everyone.

Pedagogy of identity aims to be no more or less relevant for whites than for People of Color, or for heteronormative students than for queer students, or any other binary pairing of dominant and marginalized identities. Furthermore, through exploring our own identities and experiences together, we see that despite very real structural injustices enacted in society, the binaries are false. As I have argued throughout this book, socially constructed realities are real only in their consequence. Thus, through studying identity in the paradigm of relevant culture (and multiple cultures) students can learn to appreciate both the similarities and differences in experience associated with identity in their classrooms and their communities and challenge binary conceptions of identity and oppressive ideologies.[12]

Only dominant Western culture has the privilege of marginalizing and ignoring other cultures. As such, considerations of inclusion are not transitive. "When we recognize the contingent nature of justice…the impact of identity becomes apparent, and it's here that justice becomes slippery, unstable, and controversial" (Clayton & Opotow, 2003, p. 300). As certain culturally constructed markers of identity are imposed on individuals rather than claimed, relations of power are significant in the construction of identity.

> Identity not only defines, but also facilitates certain interactions and power relationships. Because of these power differences, justice is not always transitive: fair ways for person A to treat person B are not necessarily fair ways for person B to treat person A (e.g., boss-subordinate or doctor-patient relationships. (Clayton & Opotow, 2003, p. 299)

An example of the intransitivity of justice in its relation to power and culture relates to the very real existence of racism and the imaginary existence of reverse racism[13] and affirmative action. If affirmative action sought to create more opportunities exclusively for white men, that would be extremely racist.

However, given the continuing history of overt and structural discrimination, attempts toward "equal opportunity" that reconcile the imbalances by centering People of Color are not reverse racism.

Similarly, the failure of the curriculum of dominant Western culture to speak to young People of Color does not reflect prejudice on the part of People of Color in the same way that a member of the dominant culture may assume that the curriculum of the non-dominant culture could not speak to them. The first instance represents a failure of the dominant culture to speak to people whose identities are marginalized[14] (rather than a failure to listen), and the second represents a failure of dominant culture to listen to people whose identities are marginalized (rather than a failure to speak). Failing to listen to someone does not imply that that person failed to speak and failure to speak to someone does not imply that the person hasn't listened. I made the mistake for much of my life of thinking that Hip Hop didn't speak for or about me and assuming that it didn't speak to me (as an affluent, suburban white). Because of this mistake, I did not grow to appreciate Hip Hop until recently. Not only did I rob myself through such prejudice of aesthetic pleasure, I missed what could have been an important opportunity to become more critically conscious at a younger age.

When certain elements of a unit on non-dominant popular or traditional culture do not seem to speak to a student of privileged identity, there is a significant lesson to be learned about equitable redistribution of power. I've seen white students so quickly outraged and insulted and felt they were excluded when a single lecture (or discussion) did not seem to be about them or relevant to their experience. However, the other twenty-three hours of their day are exactly about them and center on their experiences and identities. Such a shift in power is not oppression, it is not reverse racism; it is justice. Beyond the initial shock that justice may cause for whites, such inclusion has no cause to threaten whites' identities; rather, it should enhance everyone's sense of identity.

If at some point Hip Hop spoke primarily to Black youths, it has certainly grown in its appeal to youth worldwide, maybe in part because Hip Hop so often speaks to the urban[15] experience, power, and real-life struggle. That said, any pedagogy of identity, including Hip Hop, has to be founded in popular education and led by those who identify with the culture being explored. I have no right, and neither am I capable of choosing the important works of Hip Hop for study, because I am not Hip Hop. I've grown to enjoy it, but I am not it. This is a line that would be crossed with much insult in the exploration of any lived experience. One can be knowledgeable, respectful, and appreciative of an identity but never entitled to any lived experience but one's own. This doesn't mean that educators should shirk their duty as educators to develop a curriculum and search for works with which their students may connect. The educator must be a co-learner and the students must be co-teachers. As experienced scholars, educators also have the opportunity to model their own learning process. In this way, not only the content of Hip Hop but also the approach to studying Hip Hop facilitates redistribution of power in the classroom by empowering those to whom Hip Hop speaks most profoundly (often students more frequently disempowered by standard curriculum).

The educator's role

> This approach can develop rich, robust questions and understandings about specific groups, their histories, and their traditions. However, if the topic of a certain cultural group is approached merely as "adding color" to the curriculum, teachers run the risk of stumbling into any one of a number of pitfalls that run counter to the critical multicultural approach.… Such pitfalls include perpetuating stereotypes by painting a group of people with a broad brush; "exoticizing" the "other" through a shallow "tourist" approach; or, even more damaging, developing new pigeonholes by reinforcing a limited understanding of the experiences of a group of people. (Bode, 2012, p. 343)

The duty of the educator is to come to the students and begin the process "using a problem-posing approach and constructing curriculum with students on topics that both teachers and students want to explore" to create "an authentic learning experience" (Bode, 2012, p. 343). For example, an exploration of Hip Hop must focus on the works that are relevant and meaningful to students. In such a learning collaboration, new meanings and connections are formed and emergent paths of inquiry may be followed. As the learning community explores the materials together and makes connections to their own cultures, opportunities would arise to expand horizons through exploring those connections.

It is also the obligation of teachers to share their own passions with students and to reciprocate the students' contributions of meaningful pieces relevant to their own experiences within or beyond Hip Hop. In that way, a committed educator has no cause to shy away from an exploration of Hip Hop with students for lack of expertise in Hip Hop. In fact, "when the teachers announce their own curiosity and model their own struggle with ignorance, students are empowered to ask previously hushed questions and uncover misconceptions" (Bode, 2012, p. 344). However, Bode (2012) certainly did not suggest that such an inquiry process relieves educators of their obligation to prepare for class. In my own experience guiding student inquiry, I have found that it takes significantly more preparation than teaching a standard curriculum.

Praxis

This is not a theoretical idea. Hip Hop has been successfully brought into the classroom by educators such as Duncan-Andrade and Morell. Duncan-Andrade and Morell (2007) are not writing theory; they are writing praxis based on their own classroom experience and research. One fantastic unit focused on reading lyrics, books, and magazines associated with Hip Hop and popular culture along with classical works (Duncan-Andrade & Morell, 2007). The unit focused on language comprehension, critical analysis, and prominent cultural issues. Again, learning objectives were met and students were empowered. In the course of studying popular culture and current events, race and racism, power, and justice emerged as dominant themes. Students became involved in working toward change in their schools and communities. At the culmination of this unit, the students initiated a school magazine project that included articles, poems, and drawings focused on the social conditions of their school. They even interviewed local politicians (Duncan-Andrade & Morell, 2007).

In the way in which hunger was explored in Chapter 6, I use Hip Hop here as an example of engaging students with culturally relevant materials, but I do not suggest that Hip Hop is the best or only way to do that. A practice inspired by pedagogy of identity could just as well center on punk rock, world music, visual arts (from street art to "fine" art), performance, film, popular culture, or traditional culture. Likely exploring identity and engaging in activities that connect with students' sense of identity or cultural studies designed to inspire a sense of connection with other world cultures would engage students with a variety of topics.

The sense of connection to and appreciation of cultures, identities, and experiences beyond one's own allows one to see a larger, more complex, and nuanced world where one can begin to develop a sense of community and a sense of place, both locally and globally. As culture shapes the environment (especially urban) and the environment shapes culture, the process of exploring culture and identity has the potential of enlightening our ways of being in the world. At the very least, gaining an appreciation for the culture in our communities likely influences how one values one's community. I posit that developing one's self-concept through explorations of identity, culture, and connectivity fosters engagement and thus contributes to the project of restoration, in part, by establishing the foundations for one's sense of purpose.

· 8 ·

PEDAGOGY OF PURPOSE

> A musician must make music, an artist must paint, a poet must write, if he [*sic*] is to be ultimately happy. (Maslow, 1943)

This chapter identifies the importance of purpose as an element of restoration and considers how educators can help students seek purpose. The chapter begins with an exploration of what purpose is and why a sense of purpose is important. Next I suggest that purpose is not the profound and noble pursuit of the highly enlightened, but often is a more grounded and humble commitment to anything one finds meaningful. Then I explore how we develop a sense of purpose and how educators can support students on their path to purpose. The chapter concludes with a consideration of the positive youth development approach that ultimately circles back to the pedagogy of basic needs, establishing a synergy among the elements of pedagogy for restoration.

"Purpose is a stable and generalized intention to accomplish something that is at the same time meaningful to the self and consequential to the world beyond the self" (Damon, 2009, p. 33). It is a source of resilience and it provides a sense of meaning and a source of motivation, and contributes to one's identity. Additionally, when students see purpose in their education or how their education supports their own sense of purpose, they will be more invested in learning. In this way, students will be better equipped to overcome barriers to their success. For example, if students feel that a high school diploma will help them achieve their goal, they may be more motivated to get through classes that do not appear to directly serve their purpose or speak to their experience. However, we need to help students see purpose in all of their education; as discussed in Chapter 7, we have to help kids run the gauntlets as long as the gauntlets exist.[1] Despite often lacking clear attention to student identity and purpose, many core subject areas do importantly serve students later in life. For example, mathematics can certainly be taught in a way that is relevant to students' lives and serves an important purpose, but many students dislike math precisely because they struggle with the topic and are not motivated to overcome their struggle by a clear understanding of the importance of mathematics.

The fundamental need for purpose was something Kurt Hahn (1960) sought to fulfill by creating a moral equivalent to war. In a speech to the Outward Bound Trust, Hahn (1960) said "[William] James hated war but he admits that war satisfies a primitive longing of men [*sic*] which will never be extinguished, to lose yourself in a common cause, which claims the whole man [*sic*]." Hahn (1960) was inspired by his perception of despondency among youths at the time, which he attributed to the lack of a "moral equivalent to war." This is little more than a poetic sentiment and Hahn's work far surpasses any equivalency with war.

Hahn (1960) suggested that

> there are three ways of trying to win the young. There is persuasion, there is compulsion and there is attraction. You can preach at them, that is a hook without a worm; you can say, "You must volunteer," that is of the devil; and you can tell them, "You are needed." That appeal hardly ever fails.

Thus, through Outward Bound, Hahn (1960) appealed to young people's duty to serve. Outward Bound provided young people with a sense of belonging and purpose through emphasis on fitness, expedition, projects (especially service projects), and rescue services (e.g., search and rescue).

Hahn (1960) shared an inspiring story of a boy who found purpose:

> There was a wild and ruthless boy. I received daily deputations urging me that I should get rid of him. But I hesitated, wondering whether his undefeatable spirit could not be directed towards worthwhile aims. I persuaded him to join the agricultural

> guild. One day he came to me in great excitement and said, "Mr. Hahn, something terrible has happened. They want to kill our cow because she has swallowed a piece of wire. Now I know from my father, who is a doctor, that in such cases you can operate on humans, and a cow is a much tougher animal than a man or woman. May I find a vet who will operate?" In my stupidity, I said to him, "Well, if you find a Vet, who will operate on this cow, she shall live." He disappeared for twenty-four hours, then came back with a Vet who lived 30 miles away. His coming cost more than the cow. I shall never forget what happened then. I was in bed at that time with a broken leg, and the stable was 200 yards away—suddenly I heard a cry. It was the cry of a savage, but it was not the cry of fear, it was the cry of jubilant mercy. What had happened? The boy had been allowed to assist at the operation; he had poured much iodine into the wound made by the surgeon's knife. The arm of the vet, buried in the recesses of the cow had extracted the piece of wire from near the heart. When he held it up the boy emitted that triumphant cry which I had heard. He has since become a distinguished surgeon.

When students find meaning and purpose, regardless of what that meaning and purpose is, they engage with life in a way that is constructive and creative rather than destructive and they contribute to the project of restoration rather than degradation.[2]

Regardless of how and where each individual takes their passion, when humans find purpose and dedicate themselves to it, the environment (social and ecological) will take care of itself. When people are engaged and passionate about any prosocial or pro-environmental cause, it potentially benefits all causes by uplifting everyone (see the section "Prosocial Motivation" in Chapter 3). Furthermore, when people engage at any level of our collective identities (e.g., family, community, nation, or planet), they will inevitably intersect with other engaged individuals who are active in other levels of the collective. In this way, all restorative efforts are interconnected and synergistic and contribute to the broader project of restoration. If everyone is engaged in doing something to benefit life, regardless of what it is, we will see positive large-scale change. For that reason, pedagogy of purpose is about helping people to find their own sense of purpose rather than advocating any specific cause. Pedagogy of basic needs and pedagogy of identity are both aimed at helping people reach an emotional and physical place (near the top of Maslow's hierarchy) where they may discover purpose. Both pedagogy of basic needs and pedagogy of identity already facilitate the development of purpose (e.g., helping others meet their basic needs).

Purposes of Grandeur

Purpose does not have to be grandiose. Damon (2009) shared a story about the manager of a fast-food restaurant who took pride and found purpose in that work. Although one could just as easily succumb to prevalent cultural notions of such a position as a meaningless, short-term, low-wage job, the manager saw purpose and meaning in treating each transaction as an opportunity to make people smile. The manager felt that anything that they could do, however small, to make a customer's day better mattered. This sentiment rubbed off on the young adults working there as well, and the entire staff became engaged in the meaningful effort to make restaurant patrons happy. Rather than seeing their work as a meaningless, short-term job, the young employees found a sense of purpose and a source of pride in their work. Damon (2009) suggested that the manager's employees were not only positively influenced at work, but through the experience at the restaurant, their personal lives also benefited.

The manager's sense of purpose was likely independent of the job and it would probably positively influence people's lives in any position. The manager seemed to have aligned the position with what was almost certainly a previously existing desire to positively affect others (beyond any mandate of customer service professionalism). One may suggest that such purpose within certain structures perpetuates degradation. Certainly the fast-food industry is a significant producer of environmental and social degradation. A neoliberal ideologue might even suggest that the manager and restaurant patrons are complicit in such degradation (if such degradation were to be acknowledged). I see this as a conflation of distinctly separate processes. The manager clearly works, as we all do, within a problematic structure.

Who is to say that such a positive and genuine person (as the restaurant manager was described to be) would have more or less influence in any other career? In fact, as illustrated by the manager's story and certainly millions of individuals working in service industries, one can have a profound impact on others in many careers.[3] Education certainly provides the opportunity to encourage commitment to prosocial and pro-ecological causes and there is a strong tradition of such efforts in the liberal arts. However, formal education is not the only way to find meaningful, prosocial, or pro-environmental purpose. We need restaurant managers, gas station attendants, and construction workers who are dedicated to a better world (including ecological and social causes) just as much as we need teachers, academics, and politicians. We also need to collaborate more consistently across such professional divides. We need everyone to engage everywhere.

Career vs. calling and higher education

Damon (2009) described a sense of calling as a realistic awareness of one's own ability, an interest in how those abilities can be used in the service of meeting the needs of one's community (at any level from family to planet), and a feeling of enjoyment associated with using one's abilities to meet those needs. One's career and one's calling can be fully aligned or entirely independent (potentially even at odds). Furthermore, one's career does not a life make. Education needs to focus holistically on helping people live a good and meaningful life; it should not focus solely on career preparation. We place too high an emphasis on economic preparation in schools and by so heavily emphasizing the necessity of college, we derogate those who choose not to (or are unable to) attend college. This belittles the important and necessary work that the majority of people do, which does not require a college education. We need to be working to make it possible for everyone (who wants to) to attend college without devaluing those who do not attend college. Furthermore, if we concentrate on all those who are purposeful and motivated to make the world a better place in careers that are most frequently recognized as helping professions (e.g., medicine, education, and social work), we diminish the reach of social efforts. We risk losing great and important teachers, such as the fast-food manager, who reach folks that classroom teachers may not. Many of the teachers who have been influential in my own life have been those I encountered in retail, factory, and warehouse jobs that I worked at before attending college.[4]

Development of Purpose

Purpose isn't just about what people find their purpose to be but also why it matters, why they care, and how they intend to achieve their goal. Educators must be careful to focus on and support young people's own goals rather than our goals for them. Educators cannot become passive or complacent, though. Instead, we should encourage and guide students in the search for purpose. We need to help students develop and work toward their goals and offer our opinions, perspectives, and direction (how) and help them to clarify their own values and interests (why). We need to help students explore their interests through volunteer opportunities with established organizations and through helping them design and implement their own projects. We should also connect them with positive role models and mentors in the community who can offer different perspectives and opportunities, and show students what the work really looks like.

Exploration

Although purpose is stable over some period of time, it is subject to change (Damon, 2009). It may be that someone's purpose is focused on accomplishing a specific goal and upon completion a new goal will be set (e.g., raising $1,000 for a local food bank) that may or may not share a general thematic purpose (e.g., fighting hunger). For others, purpose may be a lifelong ambition. Damon (2009) emphasized the long-term benefits of a life of purpose, but shared many stories of how short-term, specific goals have led to lifelong purpose. Educators must support students in their efforts to find purpose whether it is through exploration and short-term efforts or more focused guidance in the direction of an already developed interest. Explorations and short-term efforts should be encouraged and reflected on to determine whether and how the student might explore that cause further or investigate a different interest.

Even when students become engaged with a particular pursuit, we must remain open and sympathetic to the fact that interests, passions, and purpose are not static and may shift. We need to help students stay on track to a certain extent but be prepared and willing to explore new directions with students. We should encourage and support students not to give up when they encounter barriers or abandon a cause when an initial surge of passion wanes in pursuit of some fleeting interest. However, even fleeting interests are worthy of exploration and through such exploration students may discover deeper interests or common themes among their interests and develop or clarify their sense of purpose. Additionally, this is not to suggest that we should expect every student at any grade level to discover a profound long-term purpose—that is a lifelong pursuit. Our aim should be to help students on to a path to purpose and help them develop tools to discover and develop it.

As Damon (2009) suggested, educators must "listen carefully for the spark and then fan the flames." We must allow students to discover their own sense of purpose while supporting them in the process. What is inadmissible is for educators to take students' power from them by too eagerly grabbing the wheel in an attempt to support their efforts. The balance is one between showing support by being involved, attentive, and engaged and becoming domineering through the same behaviors (Damon, 2009). Continuing Damon's (2009) metaphor, we must be very careful that we don't smother the spark in our effort to support students. One way that we extinguish the flame is by focusing too heavily on achievement. Achievement is not equal to purpose. Damon (2009) suggested that it is not the achievement that is a source of joy but the long and sustained effort toward a meaningful aim. Achievements are little more than checkpoints along the way and dedicating oneself to a purpose is about more than personal achievement. Such economic goals of personal achievement often emphasized in schools are short sighted and unconducive to deep contemplation of meaningful, lifelong purpose (Damon, 2009). In fact, by suggesting that such purposeful or purpose-building efforts as community service primarily serve the student as a measure of academic (or extracurricular) achievement and as a means for economic (or other personal) success (e.g., getting into college), we undermine the true benefits of service. Such extrinsic motivation may sap intrinsic motivation. By shifting young people's budding sense of duty from a truly altruistic effort into a discrete, one-time effort, focused on their personal gain, we rob them of what may develop into a lifelong source of meaning and fulfillment.

Supporting purpose

As explored by Jackson (1993), everything that we say and do (or don't say and don't do) around young people matters. Students often don't realize how deeply educators care about them and their success and teachers should explicitly discuss their own sense of passion, pride, and purpose in teaching[5] (Damon, 2009). By sharing our purpose, we build the student–teacher relationship, support our students, and inspire their own senses of purpose. However, supporting kids on a path to purpose is more than uncritically or disingenuously encouraging them. Students know (even if not explicitly) when adults in their lives do not genuinely believe in them and support them. We can't idly promise that they'll achieve all of their hopes and dreams (cf. Duncan-Andrade, 2009, on hokey hope and mythical hope). We have to genuinely know that students can fulfill their dreams while at the same time being honest and critical with them and avoiding meaningless platitudes.

The nature of the relationship between teacher and student significantly affects performance. Even measures of IQ, which are largely considered to be static, have been long known to vary according to the effect of the demeanor of the examiner on the emotional state of the test subject. Subjects who are treated more warmly by the examiner or have a previous relationship with the examiner score higher on IQ tests[6] (Sacks, 1952). Sacks also discussed a 1936 study finding that African American children scored lower when tested by white examiners than when tested by African American examiners and that the opposite was true for white children.

Teachers' expectations for a student become a self-fulfilling prophecy (Rosenthal & Jacobson, 1968). In one study, Rosenthal (1968) informed teachers that certain randomly selected students were "growth spurters" who were expected to be more successful in class based on a fictitious Harvard Test of Inflected Acquisition. Those students displayed greater intellectual growth than the control students (Rosenthal & Jacobson, 1968). The implications and validity of this study are hotly debated; however, it importantly raises the issue that despite our intentions, our expectations of our students almost certainly influence them. Rosenthal (1987) summarized the implications of this research by saying,

> When our children were in school we didn't want them taught by teachers who "knew" they couldn't learn. Now that our children are grown, we don't want our grandchildren taught by teachers who "know" they can't learn. Sometimes assigning some Bayesian priors seems simply prudent.

Rosenthal (1987) is suggesting that even if one challenges the statistical validity of the finding, we accept that teacher expectations have some subjective or uncertain but certainly undesirable influence on students.

Damon (2009) described how low expectations can become developmentally formative. For example, when a dream is deferred, one may interpret it as a personal failure that could lead to feelings of inadequacy that may influence one to defer or decline dreams for the future and ultimately result in hopelessness and self-defeat (Damon, 2009). It boils down to the old cliché: whether you think you can, or think you can't, you're right. Educators have a significant opportunity to positively intervene in this cycle. We have to maintain high expectations for our students and encourage them to have high expectations of themselves. Instead of deferring dreams, lowering expectations, and losing hope, we should encourage students to follow the wise mantra of the

Little Engine That Could, which has propelled me to the top of many mountains: "I think I can. I think I can. I think I can.... I thought I could."

Critical hope

"Children can survive—and thrive—in the hardest of circumstances. The children most likely to thrive are those who move towards positive goals without letting hardships deter them" (Damon, 2009, p. 171). As educators, we must help students avoid catastrophic thinking, defeatism, panic, and blame. We must focus instead on what control we do have. We need to help students gain the confidence to change the world (even in the smallest of ways).

The concept of critical hope (see the section "Hopeful Approach" in the Introduction to this book) is a composition of two distinct processes: (1) "critical social analysis of inequities and actions taken to achieve greater social justice" and (2) "a sense of personal efficacy or perceived capacity to effect change" (Christens, Collura, & Tahir, 2013). The goal is to achieve both.

> The achievement of critical consciousness...requires people to attain an understanding of structural and historical forces that impinge on and constrain collective action for social justice, yet to simultaneously maintain their hope that—despite these forces—their own participation in justice-focused action can be effective. (Christens et al., 2013)

Christens et al. (2013) identified four clusters of students based on the relation between their social critique and sense of hope and efficacy: critical and hopeful, critical and alienated, uncritical and hopeful, and uncritical and alienated. The conceptualization of these four clusters may offer some insight useful to our teaching practice. As mentioned in the Introduction to this book, it is likely (but contested) that humans are more motivated by a greater emphasis on positive solutions than identifying problems. However, differential emphasis on problems and solutions may be effective based on one's assessment of a student in consideration of the four clusters.

Positive psychology is important for everyone, and a balance must be maintained. For youths to find purpose, they must be encouraged to think positively in order to believe that they could have any effect toward their purpose. However, it seems likely that those who are critical and alienated would benefit from a greater focus on positive solutions and a lesser focus on identifying problems. But a different understanding of the problems may be more conducive to hope. Those who are extremely hopeful but uncritical may need to focus more heavily on critique in order to direct their hope toward solving real problems.

Positive Youth Development

Beyond positive expectations and critical hope we need to broadly consider the way children are viewed in Western society. Damon (2009) suggested that we have a "tendency to perceive children as fragile and inept creatures, who need constant protection from the ordinary challenges of everyday living" (p. 170). Damon (2009) suggested that "it is one thing to warn children about dangers of playing in traffic, and another thing entirely to prohibit activities that, in all but freak incidents, lead to little more than bumps and bruises" and asked, "what kinds of messages are such prohibitions sending to children about what they are able to handle?" (p. 169).

As already explored (see the section "Educational Action" in Chapter 6), young people, "if given the opportunity, can become…active participants, and powerful agents of social change" (Nygreen, Ah Kwon, & Sánchez, 2006, p. 108). Therefore, we shouldn't infantilize children by protecting them from failure or from unpleasant realities, preventing them from taking risks, or declining to assign them responsibilities. Because if we fail to treat young people as capable human beings, we engage in the self-fulfilling prophecies that Rosenthal (1968) identified.

The positive youth development approach embraces risk and opportunity and emphasizes the manifestation of young people's assets and potential over their (externally and internally) perceived deficits (Damon, 2009). "The positive youth approach aims at understanding, educating, and engaging children in productive activities rather than correcting, curing, or treating them for maladaptive tendencies or so-called disabilities" (Damon, 2009, p. 168). In accordance with such an approach, opportunities need to be facilitated in such a way that youths are inspired to become involved and initiate their own projects with an authentic sense of purpose rather than pushed into predetermined activities. Positive youth development mandates youth-led participatory action research. Thus, pedagogy for restoration comes full circle. Moral education and the three elements of pedagogy for restoration (pedagogy of basic needs, pedagogy of identity, and pedagogy of purpose) complement and reinforce one another. I have proposed that it has to start with meeting basic needs, but beyond that restoration is not a linear process. We find purpose in meeting our basic needs; we develop our identity in seeking purpose.

Finally, purpose is not a strictly individualistic concept. It has been primarily explored as such because pedagogy for restoration has been primarily conceptualized as an approach focused on the whole child. However, the chapter began with Kurt Hahn's (1960) speech as an illustration that purpose is often about being part of something larger. In fact, just as suggested in pedagogy of identity that we can form new group identities without sacrificing individual identities, we can forge group purposes that may be composed of many differing individual senses of purpose. Although I reject Hahn's (1960) thesis of a moral equivalent to war, he was correct in identifying the value of finding a common cause and devoting oneself to it entirely. We do occasionally see this when communities respond to natural disasters. As climate change becomes a more salient and life-threatening issue in a greater portion of the world, my hope is that rather than falling into chaos we will pull together for the common cause of sustaining life on Earth.

· 9 ·

INSIGHTS FOR ENVIRONMENTAL EDUCATION

> When we honor the hidden aquifer that feeds human knowing, we are more likely to develop capacity for awe, wonder, and humility that deepens rather than diminishes our knowledge. (Palmer & Zajonc, 2010, p. 22)

This chapter reconnects with the introduction of this book by proposing that the project of restoration is, in fact, a restoration of EE, congruent with its historical dictate (i.e., the Belgrade Charter and Tbilisi Declaration). Throughout this work, I have focused on the broad process of education and maintained that all education is environmental education. However, in this chapter I examine the specific field of environmental education. I begin by exploring a few significant themes in EE, which can and should inform traditional education and the project of restoration.

Beauty and Wonder

> Our consciousness must be neither poetry nor science, but a transcendence of both into a new realm of theory and practice, an artfulness that combines fancy with reason, imagination with logic, vision with technique. (Bookchin, 1982, p. 20)

We need bread but we need roses, too. All too often, it seems that bread and roses are assigned to different disciplines. Environmental education, more than traditional education, maintains an emphasis on the aesthetic. Rachel Carson and more recently David Sobel (both following a long tradition of nature writers) emphasize a connection to nature rooted in something deeper than understanding. Both Carson and Sobel have sought to foster a sense of wonder that is not explicit to pristine wilderness, but experienced in one's own community. In my interpretation, this approach is fundamentally one in which love for life encourages wonder and fuels the imagination. Certainly educators in all fields should help students experience nature and wonder, but wonder and nature are everywhere.

I'll share a brief story about how I came to love chemistry. I had always been intimidated by chemistry. I never learned it in high school, and in college, even at the introductory level, I was utterly lost within the first week of class. In fact, in high school, community college, and university I have enrolled in chemistry classes and quit within a week or two. That all changed because of one wonderful/wondrous chemistry professor. In this course, we distilled brandy from wine, generated electricity through chemical reactions, and made explosives. The professor seemed at all times (and especially in the lab) to be grinning and barely able to control their laughter. In fact, my lab partner and I once boiled over a beaker of who-knows-what on to a hotplate and filled the lab with terrible fumes. The entire lab had to evacuate and our professor laughed and laughed, and never showed any sign of anger or frustration.

It was not this single experience, but the consistent demeanor of my professor that shifted my perception of chemistry from that of a discipline of strict rules that I was intimidated by, to a vast source of wonder, curiosity, and excitement. Furthermore, I was no longer terrified of making a mistake. One time I overheard this professor say to a colleague, "It's still magic, isn't it?" It has always stuck with me that an experienced chemist with such expansive knowledge of chemistry still felt a sense of wonder even when overseeing mundane lab exercises. I think that is clearly the secret of education. If we maintain our own sense of wonder and love of knowledge, we inspire it in students.

The best lessons that I have shared have always been in situations where I was not formally the teacher, but a student or friend, and always an equal. Such lessons have taken place when a learner has displayed an interest or started making inquiries about a particular topic (that I happen to be passionate about) independently. I've found my role is more often to watch quietly and support students while they share their wonder and explain their discovery. Sometimes I may ask leading questions or provide some additional information based on their interests and suggest related areas of exploration that may be of interest. However, I don't want to demystify the world. I don't want to diminish the magic of learning to satisfy my own joy of teaching. Sometimes explicit knowledge can be a barrier to tacit knowledge whether that knowledge is about the natural environment or part of a moral induction.

I've often made the mistake of so enthusiastically sharing my explicit knowledge about a particular plant, rock, or animal that a child is observing that their own process of inquiry and sense of wonder are diminished. I consider this to be selfish teaching. Selfish teaching satisfies the educator's love of teaching at the expense of the student's passion for learning. I've frequently made the mistake of approaching a young person who was interested in a certain rock by launching into a lecture on that particular rock (e.g., "This rock is called granite; it formed deep below the Earth's surface as magma cooled and hardened. All of the different colored specs are called minerals and the slower the rock cools, the larger they can grow…." and on I go). In these cases, I generally talk until the kid loses interest and moves on to something else, probably hoping to not get caught picking up another rock. A quartz crystal can be a magical, valuable treasure if there's not a geologist around to spoil it (not that we always inevitably spoil it). When we are so quick to name, define, and discipline knowledge, we destroy wonder. The joy of learning is in feeling wonder, making inquiries, and creating meaning.

The natural sense of wonder experienced in nature may be the most powerful and meaningful element of environmental education. Both in nature and in the classroom, sharing our sense of wonder and helping students explore their own are important. Whether in nature or a classroom, we should attend to the lessons that are naturally present and waiting to be learned. We often refer to such lessons as teachable moments, but I think they'd more aptly be called natural lessons. A mentor once said to me that as environmental educators, we are not the teachers. The western red cedar, the sword fern, the deer, and all the other plants and animals are natural teachers. Our only job is to help students meet them and learn their lessons. Certainly this is an underemphasized way of learning that relates to all of education. Certainly such natural lessons are present in the classroom as well as in nature.

Systems Approach and Specific Ecological Knowledge

EE generally emphasizes a holistic and systems thinking approach to the natural environment compared with the emphasis in fields such as biology on a narrow segment of the natural world. The sort of molecular and atomic thinking that is often prevalent in scientific research must be integrated with a broader, holistic, and observational understanding of ecosystems. This is illustrated by recent developments in wilderness management efforts tending toward adaptive management practices in consideration of trophic cascades (Eisenberg, 2011). The adaptive management approach relies on both scientific knowledge and careful observation of ecosystems with a focus on the interdependence of diverse species and abiotic elements.[1]

In environmental education, we generally emphasize the basic ecological knowledge necessary to understand the natural world as a system of interdependent processes. The systems approach facilitates both explicit knowledge and tacit understanding that ecosystems are interconnected and important and likely motivates to a greater degree than specific ecological knowledge does. A key application of the moral research is that some specific ecological knowledge is beneficial to building environmental ethics, but it's probably not the most significant motivator for pro-environmental behavior. Environmental education programs generally do aim to facilitate experiences beyond the basic ecology lessons of their curricula focused on helping students connect with and grow to love the natural environment. However, environmental education must extend its systems thinking approach to more significantly include the ecosystems and social systems in which students live.

Sense of Place

Sense of place is not exclusive to environmental education; however, environmental education places a large amount of emphasis on sense of place that may inform other educational practice. Here, I explore what sense of place is, how it motivates people, and how sense of place may develop. Finally, I suggest that EE should avoid a greater emphasis on the value of natural places (i.e., pristine nature) than the places where participants live.

Definitions

Sense of place can be conceptualized as a composition of our attachment to place and the meaning we associate with place. Place attachment refers to our emotional connection to place and relates to our dependence on place and the relationship between identity and place (Kudryavtsev, Stedman, & Krasny, 2011). We may depend on a place as a setting for certain activities or as a means of survival. We may identify with place in such a way that a specific place becomes part of our self-concept. The meaning that we associate with place "is a multidimensional construct and may reflect an individual's environment, social interactions, culture, politics, economics, and esthetic perspectives,...a mix of reinforcing or contradictory personal experiences,...as well as history of places" (Kudryavtsev et al., 2011, p. 238). Individuals construct their own meanings of place and such meanings differ among individuals and peoples inhabiting the same place.

> As an illustration, one person may think of the Bronx in New York City as the birthplace of hip-hop culture, low-income housing projects, and his [*sic*] close friends, while another person may hold different place meanings for the Bronx such as her[2] [*sic*] community garden, the wildlife in waterways, environmental injustice, and her community-based organization. (Kudryavtsev et al., 2011, p. 232)

Attachments to and meaning of place are not always positive and aspects of place may conflict with individuals' self-conception or ability to meet their needs. For example, "collective memories of slavery and sharecropping may negatively influence African Americans' place attachment to wildland recreation areas...and parks" (Kudryavtsev et al., 2011, p. 238). Similarly, many Indigenous peoples' sense of place reflects deep meaning of and affiliation with place central to their identity and culture. However, the recent history of colonization of America and attempted genocide of Indigenous peoples (cf. Dunbar-Ortiz, 2014) likely influences Indigenous peoples' place affiliation and meaning. Immigrant, refugee, and diasporic peoples may also experience significant conflict regarding sense of place. The bilingual children's book *Recuerdo mis raíces y vivo mis tradiciones / Remembering My Roots and Living My Traditions* (TNL, 2003) (a fantastic result of a youth-led participatory action research project [Nygreen, Ah Kwon, & Sánchez, 2006]) discussed the experience of youths who have homes in United States and in Mexico. In the end of the story, a child reflects on leaving the family's rancho in Mexico to return to the United States and says to their sibling: "You were crying and saying,

'I don't want to go, but I don't want to stay'" (TNL, 2003, p. 26). The lesson is that we have meaningful and conflicted connections to place and sometimes to many places, similarly, a sense of place among other diasporic and Indigenous peoples.

Fostering sense of place

Numerous studies have correlated sense of place and pro-environmental behavior related to preservation and stewardship in valued environments and cleanup in response to human-caused environmental problems such as oil spills (Kudryavtsev et al., 2011). Thus, fostering a sense of place is likely to be conducive to pro-environmental behavior.

Fostering place attachment

"Place attachment can be developed through both (1) direct experiences with places, especially long-term, frequent, and positive experiences and (2) learning about places from indirect sources rather than direct contact" (Kudryavtsev et al., 2011, p. 236). In the first case, active engagement with a place tends to increase one's attachment. All of our experiences take place in place. Everything from visiting a National Park to an urban city park or community garden contributes to our sense of place. Additionally, our experiences with place include our diet, daily life in cities, and cultural events. Actively engaging with place refers not to passively experiencing place by observing it, but by interacting and manipulating place. In cities, actively engaging with place could include activities such as participating in (rather than observing) cultural events, planting gardens, and social interactions. In nature (including urban natural areas), actively engaging means getting off the trail, building forts, digging holes, swimming, building fires, feeling, smelling, and tasting the environment, and generally poking around. Such experiences both include and encourage service in natural and urban places ranging from stewardship of natural environments to community service in urban environments.

Our attachment to place is related to the way we experience place. For example, those who frequently visit an urban park feel a greater attachment to that specific park than do volunteers engaged in stewardship of an urban park who are more likely to generally feel attached to the ecosystem that the park represents (Kudryavtsev et al., 2011). Additionally, among "environmentalists," attachment to place is often associated with childhood experience related to "a sense of security, the ability to influence the environment, and the opportunity to be a functional community member" (Kudryavtsev et al., 2011, p. 237). Social interactions, experiences with animals, and aesthetic preference also contribute to one's attachment to place. Finally, through study individuals may develop a sense of place for a site to which they have never been. For example, geology students who actively studied the Grand Canyon showed some attachment to that place despite having never been there. But, their sense of place was generally weaker than that of students who had visited the Grand Canyon (Kudryavtsev et al., 2011).

Fostering place meaning

As with the two primary mechanisms by which we attach with place, there are two processes by which we assign meaning to place: "(1) creating place meanings through first-hand experiences in places and (2) learning place meanings from written, oral, and other sources, including communication with other people" (Kudryavtsev et al., 2011, p. 237). The meaning we make of place is related to the structure of our experiences and social activities.

> Experiencing unique attributes of places—including geographical features such as rivers and lakes as well as cultural attractions—may facilitate the creation of particular place meanings.... In this case the physical environment, although not deterministic of meanings, sets bounds for the possible experiences and place meanings.... For example, through experiencing an urban place it may be possible to create such place meanings as "concrete jungle" or "well-maintained public parks," but less likely "wilderness" or "native landscapes." (Kudryavtsev et al., 2011, p. 237)

Additionally, the frequency and variety of our experiences, pivotal moments, and feelings of safety influence the meaning we make of places. We may also develop meaning of local places through experiences away from those settings.

The meanings we associate with place are also socially constructed. Place meaning is "created, cultivated, and modified not only through direct place-based experiences but also by such means as stories, myths, literature, promotional materials, folklore, paintings, music, films, history, casual conversations, and memory" (Kudryavtsev et al., 2011, p. 238). Thus, narratives and conversations with others actively shape our perceptions of place. Such narratives may even be more influential than our direct experiences with place (Kudryavtsev et al., 2011). Finally, the meanings of place likely evolve over time. "For example, young people may directly experience places yet take them for granted and realize their unique ecological meanings only when they grow up" (Kudryavtsev et al., 2011, p. 238).

Environmental education and valuing places

Environmental education is quite influential in the process of developing senses of place and provides a significant opportunity for shaping participants' attachment to and meaning of place. More broadly, education should put greater emphasis on and attention toward developing a sense of place among students. However, certain practices in EE related to developing a sense of place certainly have room to grow, especially relating narratives (overt or implicit) about the value of "pristine" natural spaces such as national parks and wilderness areas. We have to work to avoid implying that pristine nature has a greater value than the places kids are already attached to and engaged with, including neighborhoods, backyards, city parks, and school grounds.

This is not to say that we should not facilitate experiences in remote or distant locations. Experiences in other environments and ecosystems, beyond one's own community, such as timberlands, agricultural areas, and even sewage treatment plants,[3] also contribute to a broader understanding of place. Certainly we should also facilitate experiences in the most pristine natural environments available. As Kahn (1999) pointed out, the environment where we grow up influences our baseline concept of nature and each generation's baseline lowers as natural environments are degraded. It is important, then, for young people to experience our exemplary natural areas such as national parks. In all cases, we should critically engage with students in discussions about land management. Additionally, as mentioned above, experiences away from those places can be important and formative for meaning related to children's sense of place in their own community. However, environmental education programs that take place in such remote or "pristine" locations must actively work to connect with and value children's own experience and continue in children's own communities.

Environmental education can and should take place in urban environments and city parks. Although urban areas are ecologically degraded, every environment on Earth is degraded, and we can't value space only in terms of the myth of pristine nature. Such a fallacy implies that any perfect environment would be ruined simply by being observed by a human. We have to recognize that a walk in an urban park can be a really huge adventure for kids if we don't let our stubborn arrogance prevent it from being so. By dismissing or overlooking children's own narrative related to wildness, we detract from the meaning of their experience. We must truly be with them and understand that they are experiencing the situation significantly differently from what we may have planned and attempt to support their experience rather than enforce our own narrative. Whether in a National Park or urban city park, we must avoid establishing a hierarchy of place where local parks and communities are lesser than the places many environmental educators live and work.

I think a story may best illustrate this point. I was recently bird watching at Deception Pass State Park in Washington State and met a family from New York City. One might suggest that Deception Pass is "better" than New York City for bird watching or for generally observing and connecting with nature. However, after a brief conversation, the child was obviously unimpressed and bored with the list of species I'd seen and said, "You should see the peregrine falcons dive from the Empire State Building." That child's eyes lit up with a sense of excitement and wonder and he commenced to rattle off what was presumably every fact that he knew about falcons. The story serves as a reminder that urban environments are rich with both natural and cultural experience and that connection with the natural world isn't about names of species and measures of biodiversity; it is about wonder and excitement for the natural world.[4] Environmental educators only have to be open to the value of such environments and adjust their narrative of nature. We can't fake it either. We can't secretly think that our "pristine" nature is better. We have to know that such environments are not lesser and feel the same wonder for the nature in urban environments as we do for the nature in national parks.

Place, value, and service learning

I recently attended a service learning project during which we made improvements to a campground. None of the kids (of those whom I was able to ask) had ever camped there. The only kids who had been there before (of those I asked) had been there the previous year to make improvements to the campground. A political representative attended to observe the program and had the audacity to proclaim how great it was to see "at-risk" youths engaged in community service efforts. These kids were indeed "at-risk." At the time, they were at risk of two things: (1) learning that community service is nothing more than free labor for the benefit of others, and (2) being derogated by the assumptions of some politician who didn't bother to get to know them and was there only for the sake of political posturing. However, these kids do struggle and cope with things that I don't think they deserve to struggle with despite their resilience. Part of that struggle is how frequently adults both openly and covertly hold low expectations for them and see them as "at-risk," disadvantaged, and hopeless.

We cannot expect those whom society has systematically silenced to raise their voice for the issues that those who have silenced them define as important. This is where many service learning efforts fail. Service needs to start in kids' own communities and address needs that kids recognize. Otherwise, the meaning of service is lost and the efforts reduce to little more than free labor. The joy of service is in the meaning and sense of purpose found in identifying a need and addressing it. The kids did potentially learn that working outside can be fun and about possibilities for future careers; however, I suspect few felt that they had solved a problem or that they would see any of the benefits of their hard work (e.g., camping there).

When we view people and places as lesser, disabled, and irreversibly degraded, we undermine our intentions to inspire folks to become actively engaged in purposeful pursuits. If the environments to be protected are "other environments" and not their own, then conservation-preservation is for the benefit of privileged others who get to live in National Parks, but not for the communities that they live in.

We certainly advocate a symbolic association with public lands (e.g., the National Parks belong to everyone) but that isn't enough. In the case of the service project, most of us opted to work for a couple of hours and spend the rest of the day enjoying nature, making art, and playing in the lake. This was a marvelous success because at least for the afternoon public land was more than symbolically theirs. However, much of the public lands in United States are no more accessible for folks in economic poverty than private attractions such as Disneyland.

Service learning must be focused on the potential that each space offers and the value that it already represents. Pristine nature does not exist and it never has existed. Certainly our consumption and exploitation of natural resources has gotten extreme, but we've rarely, if ever, intentionally left nature untrammeled. I don't think we should frame environmental issues as protecting nature from humans, but as an effort to teach ourselves how to constructively interact with the rest of life. Neither can we frame social justice as strictly focused on protecting the vulnerable. We need to focus on the assets that peoples vulnerable to oppression have and ending the systems of oppression that render them vulnerable and targeted.

Restorative Experiences in Nature

Any effort to get people outside and enjoying nature, whether in urban parks or more remote and "pristine" environments, may have intrinsic value beyond the experience. Certainly this is an assumption that EE rests on. One such intrinsic benefit of spending time in nature is that it may uniquely benefit our health and well-being. Attention Restoration Theory (ART) suggests that

> interacting with environments rich with inherently fascinating stimuli (e.g., sunsets) invoke[s] involuntary attention modestly, allowing directed-attention mechanisms a chance to replenish (Kaplan, 1995). That is, the requirement for directed attention in such environments is minimized, and attention is typically captured in a bottom-up fashion by features of the environment itself. So, the logic is that, after an interaction with natural environments, one is able to perform better on tasks that depend on directed-attention abilities. Unlike natural environments, urban environments contain bottom-up stimulation (e.g., car horns) that captures attention dramatically and additionally requires directed attention to overcome that stimulation (e.g., avoiding traffic, ignoring advertising, etc.), making urban environments less restorative. (Berman, Jonides, & Kaplan, 2008, p. 1207)

It is the environment's capacity to attract involuntary attention softly while at the same time reducing the hard capturing of voluntary or direct attention that makes it restorative (Kaplan & Berman, 2010).

Natural environments, such as parks, gardens, and lakefronts are restorative (Kaplan & Berman, 2010). They provide the necessary balance of stimulation and ability to let one's mind wonder with little that requires immediate direct attention. However, it is not strictly the qualities of an environment that make it restorative; it must be compatible with one's own interest and experience.

> Thus, if one is in a hurry to get home before it rains, an otherwise lovely creek that stands in the way of the route home would fail the compatibility test. It also helps if the environment appears to be large enough to permit one to explore it or at least imagine exploring it. Thus, a single potted plant would fail the requirement for extent. And finally, a store filled with flowering plants would not meet the requirement for "being away" if it looks just like the workplace one is seeking a break from. (Kaplan & Berman, 2010, p. 49)

However, if the experience is compatible and one is not distracted by other stressors, time in nature has restorative capacity. Berman et al. (2008)

observed that attention and memory improved when students took a walk in a peaceful arboretum (compared with a walk on city streets). They additionally found that looking at pictures of nature had restorative effects. Patients' well-being and recovery time in hospitals also improve when their rooms overlook natural settings or display images of nature (Ulrich, 1993).

It does seem that the "quality" or "authenticity" of nature does influence its restorative capacity (as far as stress reduction) despite the fact that images of nature have been found to be restorative. Kahn et al. (2008) found that recovery rates from low-stress situations (measured by heart rate) decreased when participants were able to view a natural setting through a window and that the benefit increased the longer they looked. However, using plasma screen televisions, they displayed nearly identical live footage of nature and observed no more benefit to recovery rates than a blank wall. Increased time viewing the television did not improve recovery rates. Thus, time in or observing authentic natural environments is important. However, Kahn (1999) suggested that "people may take the natural environment they encounter during childhood as the norm" (p. 2). Thus, the restorative capacity of an environment may relate to our own expectations and values of nature.

Environmental educators seem to be intrinsically aware of the restorative quality of nature and EE programs often facilitate experiences such as sit spots and silent hikes that draw on the restorative properties of nature. This is a practice that can and should inform education more generally. Whether in urban city parks, backyards, gardens, or school grounds, time in nature is an excellent way to recover from stress and fatigue. Experiencing such restorative effects of nature gives one reason to value natural areas and feel a sense of connection to the Earth.

Conclusion

Environmentalism, environmental justice, and social justice movements all grew in large part out of the civil rights movement of the 1960s and 1970s (Anguiano, Milstein, De Larkin, Chen, & Sandoval, 2012). I have argued that the needs of humanity and the needs of the planet are and must be singularly considered. However, I firmly believe in a world where many worlds fit. What I consider vital is that all people find their own passion and purpose and engage in their own way. I do not believe we will ever eliminate conflict and as I have suggested, when resolved constructively, conflict is an important source of growth. For that reason, it is important, for example, that environmental justice activists keep environmentalists in check and vice versa. Through the resolution of conflicting interests, we work toward a greater good as part of a whole in an effort for restoration.

CONCLUSION

> It is necessary to go beyond rebellious attitudes to a more radically critical and evolutionary position, which is in fact a position not simply of denouncing injustice but of announcing a new utopia. Transformation of the world implies a dialectic between two actions: denouncing the process of dehumanization and announcing the dream of a new society. (Freire, 1998, p. 74)

I found it imperative that I begin and end this project in a very personal way to communicate that this is not my book; this is myself. This work is imperfect because it is personal and in a sense it is little more than an exhaustively documented glimpse of the current stage in my own continuing development. I want to conclude by discussing the limitations of this work in an open way and in my own voice. Given the limitations and the incompleteness of this document, I reflect on the process I've been immersed in and conclude that the process has been valuable, but that I need to put what I've learned to work in a practical setting.

Limitations

This project was never value neutral. From the beginning, I have sought understanding entirely for the sake of action. I have been guided by my own sense of morality and my own vision of utopia. Thus, my investigation was hugely influenced by my own preconceptions and intuitions; confirmation bias is undeniably present in this work. With that in mind, I started with the conclusion that social and ecological degradation exists as part of a single process and sought understanding surrounding that conclusion.

The lack of neutrality is deeper than confirmation bias, though. Many of the topics of this work are deeply related to my most fundamental beliefs about life. I've always prided myself as being an extremely rational person, but throughout this process I've both discovered and developed a certain faith in life that exists beyond rationality or empiricism. I believe that there is a greater potential for life on Earth than is currently or has ever been realized. This is not an unwavering and uncompromising belief; it is something that I struggle to know because I need to know it. I wouldn't know how to live in the world that is, without faith in the world that can be. Throughout this book and throughout my life, I have been consumed with a struggle to believe that humans are good. At the outset of this book, I adopted a hopeful approach and simply refused to deviate from a position that humans are good. This bias surfaces most obviously in my considerations of moral relativism and outlying behavior. I will say that on both the topics of relativism and antisocial behavior, my own beliefs and understanding have significantly developed. But I maintain, more than ever, that people are good.

Finally, I am well aware that the thread of my thinking may be difficult to follow. I felt very strongly that the understanding I sought could not possibly be found within a single discipline, nor could it be strictly empirically or even academically explored. Thus, I've investigated in lots of directions in disciplines ranging from history to Hip Hop. I can only suggest that I think I've found several pieces to a puzzle that I haven't entirely solved; some of the pieces might not fit together and some might not even belong to this puzzle, but I feel now that I can just barely make out the picture.

A Demand for Action

As I have written this book, others have fought and died for the ideals that I tentatively defend only with words. A Zapatista educator known as Galeano was murdered by paramilitaries on May 1, 2014, for practicing the closest present-day manifestation of the ideals that I have discussed. In fact, I was in the library taking a break from reading one of Freire's books when I read the news. That day, I said to a friend that if I had any guts at all I would quit graduate school and head down to Mexico. My thought was that if nothing else, the presence of a U.S. American witness may reduce aggression. In my less dramatic moments, I have often thought that my time volunteering at after-school programs has been the most meaningful part of graduate school. At the same time, I have a constant feeling that I could spend my entire life in the library seeking a greater understanding of life.

> Because our library is…effectively infinite—no one person can ever read more than a tiny fraction—we face the paradox of abundance: Quantity undermines the quality of our engagement. With such a vast and wonderful library spread out before us, we often skim books or read just the reviews. We might already have encountered the Greatest Idea, the insight that would have transformed us had we savored it, taken it to heart, and worked it into our lives. (Haidt, 2006, p. ix)

I have had to ask myself, at what point does more knowledge and more understanding offer diminishing returns toward solving our problems? How much of my acquired knowledge simply put a name to something that I already felt, or empirically proved what I knew from life experience? How much evidence is necessary to make an effective argument that humans need to treat each other and the planet better? What did I already know, but fail to contemplate deeply? What have I failed to savor by devouring so much?

While I doubt I will ever be able to put the books down, I recognize that the time has come for action. It may be that through further academic investigation, I could extend my understanding and communicate it more clearly. However, this has never been about the writing. I do hope that this work may inform or inspire other educators, but my goal was always to inform and inspire my own action. At this point, I am satisfied with my experience researching and writing this book. In fact, I expect to continue occasionally writing and revising this document as I continue my learning process in the real world. But I know, as much as I know anything that I've learned through this process, that it is time to take a pause from studying change to go out in the world and do it.

The Unfulfilled Promise

The historical foundation of America is filled with atrocity from colonization to the genocide of Indigenous peoples and slavery and we need to face history and our roles in history as a continuing process. We have to do so to collectively heal and begin to fulfill the promise of our idealized foundation: liberty and justice for all. The Statue of Liberty is the personification of America. She stands over our eastern shore as a symbol of welcome, a beacon of freedom, and a promise of protection from imperialism. The poem, inscribed on the statue and memorized by many young U.S. Americans, says:

> Not like the brazen giant of Greek fame,
> With conquering limbs astride from land to land;
> Here at our sea-washed, sunset gates shall stand
> A mighty woman with a torch, whose flame
> Is the imprisoned lightning, and her name
> Mother of Exiles. From her beacon-hand
> Glows world-wide welcome; her mild eyes command
> The air-bridged harbor that twin cities frame.
> "Keep ancient lands, your storied pomp!" cries she
> With silent lips. "Give me your tired, your poor,
> Your huddled masses yearning to breathe free,
> The wretched refuse of your teeming shore.
> Send these, the homeless, tempest-tost to me,
> I lift my lamp beside the golden door!" (Lazarus, 2002/1903)

This was a promise that was largely fulfilled for my family and so many European immigrants, who entered United States through Ellis Island, but it is disgracefully unfulfilled for many longtime residents and recent immigrants. What welcome do those entering United States via the Mexican border receive? What New Colossus awaits them?

Aside from some insight into how and why we've failed to deliver on our promise and what an effort to reach it might look like, I haven't said much that Lazarus (2002/1903) didn't cover in those fourteen beautiful lines of positive psychology. Lazarus didn't focus on the U.S. American reality, but instead on the U.S. American potential. With equal parts rage and hope, I have tried to do the same in this work. I want to end by sharing one of Dr. Martin Luther King Jr.'s final speeches before his assassination in 1968 as he organized the Poor People's Campaign:

> In a few weeks some of us are coming to Washington to see if the will is still alive... in this nation.... Yes, we are going to bring the tired, the poor, the huddled masses.... We are going to bring those who have come to feel that life is a long and desolate corridor with no exit signs. We are going to bring the children and adults and old people who have never seen a doctor or dentist....
>
> We are not coming to engage in any historic gesture. We are not coming to tear up Washington.... We are coming to ask America to be true to the huge promissory note that it signed years ago....
>
> We are coming...to engage in dramatic nonviolent action, to call attention to the gulf between the promise and fulfillment; to make the invisible visible. (as cited in Smiley & Ritz, 2014, pp. 226–227)

We are coming

NOTES

Preface

1. The use of this term is intended to contextualize the notions of science within the academy and the dominant American culture and makes implicit the fact that there are alternate systems of knowledge. To further contextualize WMS, Pomeroy (1994) suggested a different interpretation of the acronym: White Male Science. This term is used to emphasize the role of patriarchy and white supremacy within the institution of Western Modern Science.
2. I don't assume any Black person to be, or to identify as, African American (cf. Dead Prez, "I'm a African," on *Let's Get Free*, 2000). There may be an implication of citizenship and nationalism. The term "People of Color" seems much less assuming, but I will always defer to the terms of identity, status, and experience expressed by the authors whom I cite.
3. I have made a deliberate attempt to use inclusive language in this work. I have quoted several historical texts that use archaic language in which the terms "man" or "mankind" are used in a plural sense that is not gender inclusive. In each case, I have used the Latin adverb *sic* to indicate that it is archaic language quoted exactly as it was written. Additionally, I have decided not to directly quote any text using such language if it was written within the past approximately thirty years because the contemporary use of such language is inexcusable beyond the second wave of feminism. Furthermore, I believe that even when using inclusive language, when gender and other markers of identity are not made explicit, the white male voice dominating academia may be implicit. While generally avoiding gender-specific pronouns, I have decided to directly call attention to the claimed

identities of some authors in some contexts. I do so at the risk of causing their knowledge to be interpreted or valued differently. This is absolutely not my intention and I hope that this decision does not result in a sense of tokenization of those authors. I deeply apologize if I have, at any point, misidentified any of the folks mentioned in this book or if I have used language that is exclusive.

4. I believe that the language that we use is important. I prefer the term "Jewish people" to "Jews." "Jew" is not considered a bad word (except when it is used as a negative attribute, e.g., "Jew lawyer," or as a verb, e.g., "Jew the price down"). However, as part of an effort to use inclusive language, I find it important to use life-affirming language. The term "Jewish people" affirms life and resists dehumanization. It's easier to remain disconnected when discussing "destroying Jews" because nothing about the phrase affirms life. It could just as well be replaced with "destroying furniture." A phrase such as "murdering Jewish people" makes the situation much more salient because the terms are human.
5. This is more aspirational than practical. One shortcoming of this method is that I have primarily selected my own sources. Every source is filtered through my own perceptual framework and shifted to fit my own interests and I have un/intentionally omitted ideas that have not resonated with me. My own internal conversation could never rival the diversity of thought offered in a genuine conversation with a distinct other. Ultimately, I do not possess anywhere near the cognitive prowess and theory of mind to truly reflect the voices of the many brilliant thinkers whom I have cited.

Introduction

1. For an account of the human impact on nature see Steffen, Grinevald, Crutzen, and McNeill's (2011) paper proposing that the human impact on Earth is so great that it rivals geologic forces and represents a new epoch in Earth's history that should be formally recognized as the Anthropocene.
2. "Whenever we love justice and stand on the side of justice we refuse simplistic binaries. We refuse to allow either/or thinking to cloud our judgment. We embrace the logic of both/and. We acknowledge the limits of what we know" (hooks, 2003, p. 10).

Part I: The Status Quo

Environmental Education

1. A commonly used definition of EE that is attributed to the Tbilisi Declaration is:

   > Environmental education is a learning process that increases people's knowledge and awareness about the environment and associated challenges, develops the necessary skills and expertise to address the challenges, and fosters attitudes, motivations, and commitments to make informed decisions and take responsible action. ("Environmental Education Framework: Creating an Environment to Educate About the Environment," 2003)

Although I do not suggest that this is a misrepresentation of the Tbilisi Conference proceedings, it is not a complete representation and neither is it a direct quote, as it is sometimes presented.

2. Kahn (1999) specified that the majority of the participants in the study preferred to be referred to as Black.
3. I don't condone using language of sexual violence to describe exploitation of the Earth. I think it is hyperbolic and contributes to a culture of rape. However, the nature of our affiliation with nature isn't always a loving one. What is the significance that "raping the Earth" is such a common phrase? In the next chapter, I examine Bookchin's (1982) hypothesis that the subjugation of the Earth follows the subjugation of women.
4. Grande (2004) prefaces *Red Pedagogy* with the following:

   > I am a Quechua woman. This is not only who I am but also, in these "postcolonial" times, an identity I feel increasingly obligated to claim. At the time of Spanish invasion, the Indigenous population of Peru was approximately twelve million but after the first one hundred years of conquest, numbers dwindled to 150,000—a decimation of nearly 80 percent. Despite centuries of genocide, Peru today remains a majority Indigenous population, a statistic that speaks more to the fortitude and resilience of Quechua and Aymara peoples than to the virtues of colonialist regimes. This is evident in the fact that although Indigenous peoples continue to comprise the majority, the historical-material spoils of imperialism (economic and political power) belong to the white and mestizo minority. (p. ix)
5. Grande (2004) often uses the term "whitestream" rather than "mainstream" or "Western." Although she does not provide an explicit definition of the term, it effectively centers Indigenous thought and experience by denying the assumption that dominant, mainstream, or Western ways of knowing and being include or represent Indigenous knowledge and experience.
6. Indigenous peoples of the land now occupied by the United States and more broadly the Western Hemisphere are anything but homogeneous and it is important to avoid overgeneralization. Although vast trade networks facilitated cultural and material exchange, development of cities, agricultural systems, and governance varied significantly according to region. As in all places, pre-colonial powers rose and fell and new societies emerged (Dunbar-Ortiz, 2014). However, no force before colonization ever threatened the sovereignty of the peoples to such a great extent.

Oppression

1. APA guidelines dictate that racial/ethnic identifiers be capitalized including White and Black (*Publication Manual of the American Psychological Association*, 2010). However, I intentionally challenge this precedent by capitalizing racial/ethnic identifiers of People of Color but not whites. As a result of white supremacy, whites rarely identify accordingly with skin color (a notable exception may be White Supremacists) and have the privilege to racially and ethnically self-identify to a much greater degree than People of Color and thus may identify as Jewish, European American, Irish American, etc. (Visconti, 2009), while many diasporic Peoples of Color in America do not know their

ethnicities so specifically because of colonization and the enslavement of Black peoples (Potter, 1995).

Potter (1995) suggested that capitalization essentializes and homogenizes Peoples of Color. This is absolutely not my intention; my intention is to address oppression and call attention to the prevalence of culturally imposed markers of identity in a racialized society by avoiding colorblindness. My intent is to challenge rather than essentialize racial determination.

Finally, I use the word "whites" in contradistinction to my earlier notes on using life-affirming language. I do this because (1) it makes me uncomfortable to be categorically referred to (as a white person) in such a way while similar language is rarely questioned when categorically referring to "Hispanics" or "Blacks." That discomfort is good and important and healthy and reminds me to use language carefully. (2) Oppression and racism are unilaterally enacted by the dominant group (whites), so the rights and humanity of whites are never categorically violated in the way that the rights of People of Color are. Thus, every time I use the word "whites" in contradistinction to life-affirming terms, I intend it to serve as a reminder of the nature and presence of oppression in life and in language.

2. This is intentionally left open ended and the fundamental mythologies of American civilization are potentially an endless pursuit. However, I will briefly introduce some of the complexity surrounding the myth of pristine nature as it relates to oppression. Early Western concepts of the wilderness or the wild often regarded it as dangerous and undesirable. Throughout the colonization of America, concepts of wilderness contributed to the myth that America was a wild frontier to be settled. In reality, America was not a wilderness and European and American settlers relied on the developed networks of trails and managed grasslands to establish settlements. The "wilderness" that settlers did encounter likely resulted from their own role in disabling the Indigenous peoples from tending to the land. The policy of settling wilderness was later formalized in 1823 through the doctrine of discovery, which ruled that all land lay claim to the first Christian Europeans to explore and settle it. This concept of wilderness was fundamental to the colonization of America and genocide of native peoples. Additionally, the myth of pristine nature continues to be a source of justification for the continued dispossession of Indigenous peoples of lands promised by treaties following the invasion of America. This is now done in the name of preserving wilderness rather than conquering it. Consider the dispute between the sovereign Republic of Lakota and the United States over the Black Hills (the site of Mt. Rushmore), which was established in violation of the 1868 Treaty of Fort Laramie. The U.S. government currently holds more than $1 billion in trust as restitution; however, the Lakota refuse to accept the payment because the land was never for sale (Dunbar-Ortiz, 2014).

3. Grande (2004) uses the term "indigenista" in contradistinction to whitestream feminism:

   Though Indigenous women share with other women a position of marginality and the experience of structural subordination, I believe their distinct subjectivity as colonized peoples and members of "domestic dependent nations" places the historical materiality of their lives more on a par with Indigenous men than any other sub-category of "women." I do, however, recognize the salience of gender as a category as well as the importance of a gendered, pro-woman, antisexist analysis. (p. 156)

4. Bookchin (1982) does not use the term "man" in the archaic and plural sense, but asserts that "the split between humanity and nature has been precisely the work of the male" (p. 18).

Part II: Empathy and Morality

Empathy

1. Laland & Brown (2011) discuss notes written in Darwin's own posthumously published journals in which he reminded himself to avoid referring to higher or lower forms of life. By their accounts, Darwin was very concerned about potential misinterpretation and misuse of his theories.
2. Definitions of speciation are quite controversial. The commonly held definitions of speciation is that cross-breeding is either impossible, sterile, or results in sterile offspring (e.g., mules) (McWorter, 2010). As race was beginning to be scientifically formalized in United States, it was proposed in 1943 that humans of mixed race suffered from health problems and were generally infertile (McWorter, 2010). As discussed in the section on the evolution of empathy, the use of concepts of species to establish races among humans is a distortion of genetics as a tool for racial oppression. As a final note, McWorter (2010) notes that the commonly held definition of "species" (based in sexual reproduction) is frequently used to derogate non-heterosexual and transgender individuals.

Morality

1. As discussed in the Introduction and in the context of the evolution of empathy, I find it important to know the history of ideas. I would love to have explored the history of moral philosophy, moral psychology, and human development more fully in order to understand how our understanding of these topics may have developed in social and historical context. However, these are timeless questions, and the body of work is too vast to even adequately cover current understandings. With this in mind, I have tried to at least mention some of the major influences of the scholars I have cited.
2. As the author of the poem cited by Glasser (2007) did. Recall that he was contemplating suicide.
3. Gibbs (2014), in accord with neo-Piagetian thought, does not consider such cognitive development to limit children's capacities, but instead to influence their tendencies.
4. Moral relativism is considered apart from the moral development process in Chapter 5.

5. Maybe anarchists most profoundly assert the right to freedom. However, although anarchism includes diverse schools of thought, all but the most individualistic concepts seek harmony and mature reciprocity. Thus, the personal right to freedom is limited by moral consideration of others.
6. Most people probably do not consider moral principles explicitly or discretely, but such principles likely make up a holistic concept of morality.
7. Oliner and Oliner (1988) focused heavily on the role of parenting in personality development as discussed in the previous section. The ethic of care is associated with, but not exclusive to, extensive personalities. Oliner and Oliner do investigate normative sources of motivation such as religion and culture. However, Oliner and Oliner did not significantly investigate the role of one's knowledge of family history. Recall that I found myself significantly motivated by my family history (specifically the Holocaust). My intuition is that many committed to practical and theoretical work for justice are motivated by knowledge of family history. This may be worthy of deeper investigation, but I have not pursued it in this book.

Social Influence and Individual Difference

1. I want to be absolutely clear that I do not approve of Barnett's stance on segregation. Neither do I accept an entirely relativistic stance wherein his actions were permissible according to cultural and historical context. Moral relativism is explored later in this chapter. For now, suffice it to say that I think it is possible that Barnett acted morally and that what he did was wrong. I don't consider moral action to be universally right. I consider it to be action in accordance with one's own moral judgment (i.e., the belief that it was the right thing to do).
2. In a letter published posthumously, Freire (2004) significantly deviated from his humanist tradition and may have been beginning to formulate a more holistic view of oppression. Freire wrote on the day of a brutal murder of a Pataxó Indian,

   > The fact in itself that this tragic transgression of ethics has taken place warns us how urgent it is that we fight for more fundamental ethical principles, such as respect for the life of human beings, the life of other animals, of birds, and for the life of rivers and forests. I do not believe in love among women and men, among human beings, if we do not become capable of loving the world. Ecology has gained tremendous importance at the end of this century. It must be present in any educational practice of a radical, critical, and liberating nature. (2004, p. 47)
3. There are obviously certain differences between humans and other animals that dictate that human moral considerations do not apply in precisely the same way to other animals. One significant factor is our very limited ability to communicate with other species.
4. *DSM-5* no longer diagnoses Asperger's disorder and those previously diagnosed with Asperger's disorder are now diagnosed with autism spectrum disorder (American Psychiatric Association, 2013). Autism spectrum disorder may be specified with or without accompanying intellectual impairment and with or without accompanying language impairment. Severity is further specified in relation to the level of support that the individual needs. (2013)

5. I found the investigation of trauma to be the most difficult portion of this project. The research was heartbreaking and it was difficult to maintain hope. In that sense, I mean that practicing positive psychology is as important in studying trauma as it is in treating trauma.
6. Poverty is distributive injustice justified in the minds of the privileged through defensive thinking, denial, and other moral and empathic bias. However, areas where economic poverty is prevalent are not necessarily populated by impoverished people. Those who are economically poor may be especially vulnerable to the atrocities of capitalism and lack economic, social, and political agency. In the worst cases, the most destitute are not able or barely able to meet their basic survival needs. However, the targets of such oppression are resilient and may be extremely wealthy by noneconomic measures. Furthermore, while I have tried to illuminate problems associated with economic poverty, it is vital to consider these problems as more significantly structural than individual issues. People who are vulnerable to or targeted by oppression are not necessarily living a life of perpetual unhappiness and suffering. Certainly some folks face major hardships that may, in extreme cases, be all encompassing, but others live happy and fulfilling lives in material poverty. In fact, external conditions beyond survival needs play a minimal role in happiness (Haidt, 2006). But happiness despite material poverty doesn't provide justification for the injustice resulting in poverty. What is necessary is a nuanced conception of poverty in which we neither conceptualize people as victims nor justify their oppression, but recognize injustice and make reparations.
7. Haidt (2006) uses the metaphor of a "rider on an elephant" for human behavior. The rider is our conscious brain and the elephant is the rest of our self. The rider isn't in complete control of the elephant's actions, only parts and only sometimes, and ultimately must defer to the much more powerful elephant. The elephant is composed of autonomous systems such as breathing and our beating heart as well as emotions. Confabulation means that our conscious mind explains the elephant's actions without actually understanding the elephant. Our brain is so good at this that we don't necessarily know that we've fibbed to ourselves. We ultimately accept the first plausible explanation that our mind offers.
8. bell hooks (2003) wrote:

    > Coming to academia thinking of myself first and foremost as an artist (a poet, a painter, a writer), I pursued a teaching career as an avocation. My desire was to create art. (p. 185)

9. The necessity for social change as a survival need is further discussed in the section "Meeting Basic Needs Mandates Social Change" in Chapter 6.
10. Oliner and Oliner (1988) posited that extensivity was more conducive to prosocial behavior than constriction (see the section "Influence of parenting" in Chapter 4 Influence of parenting). Einolf (2010) tested the hypothesis for prosocial behavior such as donating to charity and volunteering and supported the Oliners' (1988) claim. Both researchers explicitly state that constricted persons can or do behave prosocially and altruistically.
11. Note that this sentiment aligns with the concept of elevation as discussed in the section "Prosocial motivation" in Chapter 3. Elevation was not a theme explored in Oliner and Oliner's (1988) work. This speaks to the relationship between theory and research (Greenwald, Pratkanis, Leippe, & Baumgardner, 1986) because Oliner and Oliner aligned

their findings with the prevalent theory of their time. I hypothesize that reanalysis of the Oliners' data may reveal elevation among rescuers as a significant factor in rescue.

Part III: Pedagogy for Restoration

1. This is not intended to be a condemnation of Marxism, but of essentializing/fetishizing Marxism to the point that we impose it on students. Marxism is a fine ideology with broad interpretation in many contexts, many of which could certainly enrich education.
2. One could suggest that I am simply arguing for my own pluralistic ideology of restoration. That person would probably be right. Just as I prescribe to others, I struggle to broaden my perspectives and think beyond my own ideology.
3. *Fight the Power: Rap, Race, and Reality* (Chuck D & Jah, 1997) is one of the most influential books I have ever read. In plain language and rooted in the culture of Hip Hop, Chuck D of Public Enemy explained race in America in a way that is approachable for folks who don't use terms like "microagressions" and "structural oppression." In fact, until I read Chuck D's book almost a decade ago, I had never sought out or understood the work of most of the authors I have cited here. In fact, I didn't read Chuck D's book to learn about racism; I read it because it had a catchy title and because of Chuck D's status and visibility in pop culture. I was resistant to some of the ideas in the book, but by the time I had finished it, my entire worldview was beginning to shift. I think this exactly illustrates the role of public intellectuals—to make knowledge accessible to a community broader than the one already holding that knowledge.
4. This is not to suggest that curriculum theory, learning theory, and pedagogy are not important. Theory certainly informs instructional methods that may optimize student learning.
5. However, exploration of near-universal rights that have been established by consensus, such as the *Universal Declaration of Human Rights* (United Nations, General Assembly, 1948) and other international agreements on human rights, the treatment of animals, and environmental protection, would likely be excellent additions to complement more inductive and experientially focused curriculum.
6. Bode (2012) provided three wonderful examples of critical multicultural curriculum including Cambodia and the Cambodian American experience, expanding definitions of family, and gay and lesbian literature. Each unit incorporates multiple content areas such as math, science, English, and history.

Pedagogy of Basic Needs

1. Throughout this book I have tried to emphasize non-academic ways of knowing. Abraham Maslow's theory of motivation is an important and valuable academic contribution, but I want to recognize that his work was inspired and heavily influenced by Indigenous knowledge. Although it is sparsely mentioned in Maslow's work, Narcisse Blood and Ryan Heavy Head (2007) have spoken about it at length. Maslow spent six weeks at Siksika in 1938 with Blackfoot people. His observations of their way of life informed his hierarchy of needs and concept of self-actualization. In fact, the hierarchy of needs is thought to

have been inspired by Siksika tipi designs (Heavy Head & Blood, 2007). However, one of the fundamental needs that the Blackfeet recognized which Maslow did not was a sense of place. In fact, Heavy Head and Blood (2007) joked that Maslow's tipi was floating in space.

2. Guilt, as Hoffman (2000) defined it, is a prosocial motive encouraging a sense of responsibility or duty. Guilt is not necessarily a "bad," painful, or negative feeling unless one feels that one has failed to reconcile the situation.
3. Throughout the book I have used [*sic*] when a quote uses male pronouns to refer to humans generally. However, some authors alternate between he/him/his pronouns and she/her/hers. In each case, I wonder how and why they chose. Most likely, in the above quote and most instances it was not something significantly contemplated. One might even suggest that the choice was random. But unless the author literally flipped a coin, humans are not capable of thinking randomly. Although, not deeply explored here, I have frequently noticed that by alternating pronouns in attempt to be inclusive, gendered traits and values are subtly reinforced. Additionally, use of binary pronouns remains uninclusive to non-binary and transgender identities. In the quote, the use of she/her pronouns subtly derogates women by suggesting that although all humans confabulate, being emotional is feminine and irrational (therefore devalued).
4. There are many named approaches to participatory research and community action (e.g., Community Based Participatory Research [Anguiano, Milstein, De Larkin, Chen, & Sandoval, 2012], Action Research, Youth-led Participatory Action Research [YPAR]) and the methodological differences are sometimes quite subtle and nuanced. I take a broad approach to the concept of action research most closely following the ARCPS model because ARCPS explicitly focuses on student involvement in community problem solving as an educational process. However, other models that more broadly focus on community involvement in research and evaluation provide additional insight into the process. Many papers, including Wals et al. (1990), provide specific methods and adaptations particular to the intended setting (including environmental education).
5. I don't mean to suggest that we only work in our own neighborhoods. I mean to suggest that rather than viewing places and peoples as "others," we recognize that we are part of one large community of life composed of many overlapping communities. In this way our work is always with and never strictly for.
6. Even those acting purely altruistically without any expectation of external reward benefit internally through such experiences. It is not selfish or counter to the ideals of altruism to recognize that helping others benefits one's self.
7. In some schools, kids have their lunches taken away at checkout if they are behind on payments, surely inflicting both shame and hunger. For this manager to refuse to do this to kids at her own personal expense shows how important everyone's role is in the project of restoration. It's not the work of academics and educators; it's the work of all of us.

Pedagogy of Identity

1. Certainly in some circumstances the relative significance of particular biologically dictated and socially constructed identities change. For example, although I have only recently

considered my Jewish identity as it intersects with my other identities and influences how I experience the world, when neo-Nazis enter the equation, I am acutely aware of the significance of this particular facet of my identity.

2. Although I think that much of what Chuck D (1997) wrote applies more generally, I have tried to be true to his words in an effort to honor the voices of the authors whom I have cited rather than appropriate their ideas for my own agenda.
3. Kaplan and Berman (2010) suggested that coping with negative stereotypes and devoting attention to avoid confirming them may deplete one's resources for self-regulation of executive functioning. This pressure may be exacerbated for celebrities of color who are constantly in the spotlight. Thus, while Black public figures certainly serve as role models they cannot be expected to be perfect and we should be sympathetic to the stressors they face.
4. It is important to distinguish the public persona of artists and performers from the works that they produce. Although it is important to pay attention to the way People of Color are portrayed in film, television, and music, the consumers of the products need to be accountable for the meaning of it.
5. This is not to suggest that one (myself included) cannot or should not maintain positive associations with one's status and identity. However, there privilege associated with dominant identities is unjust.
6. By saviors I mean whites who believe that they know best and are doing anti-racist work for People of Color, rather than with People of Color. However, this mentality may well apply to folks who rescued my family from the Nazi death camps and I don't think such a mentality should be entirely dismissed or criticized. On the opposite end of the spectrum are allies. Ally is a loaded term, and I don't intend to delve deeply into different perceptions of what ally-ship means. Here, I am challenging the perception that whites can only act with or under the direction of People of Color. I cherish opportunities to act as an ally, but I think that we all need to be sufficiently informed and empowered to take immediate and independent action when it is necessary and to lead when appropriate. White co-conspirators must challenge the role of their own privilege in such behaviors without renouncing their agency entirely. I conceptualize the co-conspirator as the midpoint in a spectrum with ally and savior on either end.
7. Northeastern New Mexico was colonized by the French and purchased by United States in 1803, while the rest of New Mexico was occupied following the U.S.-Mexican War. The Spanish occupation of New Mexico swept north of the Rio Grande beginning in 1598, decimating approximately 75% of the Pueblos of New Mexico over two decades (Dunbar-Ortiz, 2014). However, in 1680 the New Mexico Pueblos organized a revolution with the help of unconquered Navajos, Apaches, Utes, and Hopis from the West and subjugated Indigenous and mestizos in the Spanish capital in Santa Fe. The revolution was successful and they drove the Spanish out of New Mexico for 12 years. The Spanish did eventually occupy the land for another 130 years until Mexico's independence in 1821. Following independence, many Indigenous peoples found refuge from U.S. genocide and relocation in the Republic of Mexico. However, Mexico allowed slave-owning Anglo-American immigrants to establish plantations, which led to conflict when slavery was abolished in 1829, ultimately resulting in the secession of the Texas Republic and

setting forth the U.S.-Mexican War. By 1848, a large portion of Mexico, including New Mexico, was under U.S. rule and relocation and genocide of Indigenous peoples increased up to and following the U.S. Civil War. By the 1890s, most surviving Indigenous refugees were confined to federal reservations and their children sent to boarding schools. However, the struggle for Indigenous rights and sovereignty continues in New Mexico and in all of America to this day (Dunbar-Ortiz, 2014).

8. Not a label the folks in this study identified with.
9. Posole is a traditional Mexican soup made from hominy that is very popular in New Mexico. It is a dish that was eaten in pre-colonial Mexico but there is a distinctly New Mexican recipe for posole influenced by Spanish and American colonization.
10. The EZLN remains an organization for and by Indigenous peoples collectively led by the Clandestine Revolutionary Indigenous Committee and Marcos has stepped down as spokesperson in order to allow for an Indigenous person to fill that role. However, Marcos remains an active member of the collective and has taken the name Galeano in honor of the murdered Zapatista educator.
11. This could be an excellent youth-led participatory action research project. In fact, I've heard middle school students clearly articulate the Eurocentric curriculum in American schools.
12. I want to note that discovering, challenging, and resolving conflicts related to strict conceptions of identity is not innately so "intense." Sometimes it's just the culmination of multiple subtle experiences or a few lyrics from a Hip Hop track. The big challenges and breakthroughs are important, but I want to emphasize pedagogy for restoration is meant to be challenging as well as fun and empowering and beautiful. I firmly believe that if it weren't, there would be no point.
13. Racism is unilaterally enacted by the dominant group. Spring (2014) provides a very simple definition of racism: "prejudice plus power" (p. 166). This definition distinguishes between simple feelings of hostility and prejudice toward another racial group and the ability to turn those feelings into some form of oppression. However, both enactors and targets of oppression internalize racist ideology and perpetuate it. For a humorous explanation of the distinction between racism and "reverse racism," see Aamer Rahman's routine on reverse racism (Rahman, n.d.)
14. Significantly, the identities of People of Color are frequently marginalized. However, this failure in education does not apply exclusively to People of Color; many identities are institutionally marginalized. Additionally, I want to be clear that no one's identity is innately marginal (see the section "Oppressor-Oppressed Dichotomy" in Chapter 2).
15. "Urban" is a bit of a vague and euphemistic term that is often used to dodge issues of race, poverty, and class. Nygreen, Ah Kwon, and Sánchez (2006) challenged this "dominant discourse that frames urban youth as disengaged or troubled" and suggested "urban" is "a euphemism for underserved, poor, marginalized, ethnic minority" peoples (p. 108). I have avoided using the term as a label of identity (e.g., urban youth) in an attempt not to be presumptuous about the identities of the referents of the term because urban refers to a type of environment. For that reason, I do discuss urban environments and the experiences that may be common urban environments (e.g., "the urban experience"). This is still a euphemistic term that is loaded with assumptions and generalizations related to race

and socio-economic status, but I attempt to avoid explicitly making such assumptions and use the term "urban" for lack of a good substitute.

Pedagogy of Purpose

1. I want to be clear about my position on the gauntlets and the oppression perpetuated in schools: public education, can, will, and does happen with or without schools (cf. Illich, 1970/2011). As outlined by Illich (1970/2011), in a society that values children and education above all, schools would be unnecessary. However, I want to make it clear that I am absolutely and unquestionably a proponent of public education. I see all of the elements of pedagogy for restoration as both radical and pragmatic. In other words, all of the suggestions that I have made are possible within existing public schools or as an addendum to the process through community efforts such as after-school programs. Unquestionably, my dream is to establish a fully democratic public education system free from the problematic structures of the existing public education system, but until that is possible, I think we have no choice but to work within the existing systems.
2. Antisocial purposes also exist both in the context of "evil" and in complicity with social injustice. Certainly Manifest Destiny provided a sense of purpose to American settlers. This is certainly worthy of further exploration in light of human morality, but I will not be dipping back into deviance or outlying behavior in this discussion. Damon (2009) provides some exploration of deviant purpose.
3. Additionally, such direct social interaction is not the only way positive contributions are made to society. Our entire society rests on the shoulders of those working in service industries. Our lives and dedication to "higher" purposes are made possible by millions of people who work to ensure that we are able to do things like buy food in the grocery store.
4. The point is that great educators (or otherwise positively influential people) are found in all facets of life and that attempts to concentrate such people into specific fields would diminish opportunities for education. I hope that it is implicit, but I will make explicit the fact that I also recognize and appreciate the profound and important role of all of the formal educators in my life from elementary school to community college to graduate school.
5. Those who cannot genuinely express such a sentiment may need to reassess their own sense of purpose as an educator.
6. Although this example is intended to highlight the importance of the student–teacher relationship, other considerations of education bias on intelligence tests and cultural bias on standardized tests shouldn't be ignored.

Insights for Environmental Education

1. Indigenous peoples in America have understood and practiced their own form of adaptive management for thousands of years. However, "management" may not be the most appropriate term for the relationship between the Indigenous peoples and the land. Native peoples didn't make the same distinction that Europeans did between wild and domestic. Nevertheless, there was likely no wild/unmanaged land in what is now United States by the time European settlers arrived (Dunbar-Ortiz, 2014).
2. See footnote 9, Chapter 6, for a discussion on use of alternating pronouns. Here, I will simply raise the question: What implications is this quote making about gender roles and social expectations of women, men, and non-binary or transgender identities?
3. Sewage treatment facilities represent fantastically unique ecosystems. Additionally, one of my community college environmental science professors used to say that all citizens should know where their community's resources come from (e.g., water, food, electricity) and where their waste goes (e.g., landfills, sewage treatment). These are important elements of human ecosystems.
4. I do not mean to suggest that conservation and preservation efforts are unnecessary because one can observe nature in urban environments. Instead, the underlying assumption is that helping people value and connect with local environments would encourage a broader concern for the natural world, especially when paired with basic ecological knowledge. If one grows to love peregrine falcons in New York City and learns about the species' migration pattern and how DDT historically endangered the falcons, they may become passionate about preservation and conservation efforts.

REFERENCES

Abrams, I. R., Coast Community College District, & KOCE-TV. (2002). *Faces of culture*. Olathe, KS: ACT/RMI.

Alisic, E., Jongmans, M. J., van Wesel, F., Kleber, R. J., Alisic, E., Jongmans, M. J.,…Kleber, R. J. (2011). Building child trauma theory from longitudinal studies: A meta-analysis. *Clinical Psychology Review*, *31*(5), 736–747. doi:10.1016/j.cpr.2011.03.001.

American Psychiatric Association. (2013). *Diagnostic and statistical manual of mental disorders: DSM-5* (5th ed.). Arlington, VA: Author.

Anguiano, C., Milstein, T., De Larkin, I., Chen, Y.-W., & Sandoval, J. (2012). Connecting community voices: Using a Latino/a critical race theory lens on environmental justice advocacy. *Journal of International and Intercultural Communication*, *5*(2), 124–143. doi:10.1080/17513057.2012.661445.

Austin, W., & Hatfield, E. (1980). Equity theory, power, and social justice. In G. Mikula (Ed.), *Justice and social interaction: Experimental and theoretical contributions, from psychological research*. New York, NY: Springer-Verlag.

Baszile, D. T. (2009). Deal with it we must: Education, social justice, and the curriculum of hip hop culture. *Equity & Excellence in Education*, *42*(1), 6–19. doi:10.1080/10665680802594576

Berenguer, J. (2007). The effect of empathy in proenvironmental attitudes and behaviors. *Environment and Behavior*, *39*(2), 269–283.

Berenguer, J. (2010). The effect of empathy in environmental moral reasoning. *Environment and Behavior*, *42*(1), 110–134.

Berman, M. G., Jonides, J., & Kaplan, S. (2008). The cognitive benefits of interacting with nature. *Psychological Science, 19*(12), 1207–1212. doi:10.1111/j.1467-9280.2008.02225.x.

Berman, S. (1997). *Children's social consciousness and the development of social responsibility*. Albany, NY: SUNY Press.

Bode, P. (2012). Adapting curriculum for multicultural classrooms. In S. Nieto & P. Bode (Eds.), *Affirming diversity: The sociopolitical context of multicultural education* (6th ed., pp. 342–377). Boston, MA: Pearson Education.

Bookchin, M. (1982). *The ecology of freedom: The emergence and dissolution of hierarchy*. Palo Alto, CA: Cheshire Books.

Brown, N. J., Sokal, A. D., & Friedman, H. L. (2013). The complex dynamics of wishful thinking: The critical positivity ratio. *American Psychologist, 68*(9), 801–813. doi:10.1037/a0032850.

Cacho, L. M. (2012). *Social death: Racialized rightlessness and the criminalization of the unprotected*. New York, NY: New York University Press.

Call, J., & Tomasello, M. (2008). Does the chimpanzee have a theory of mind? 30 years later. *Trends in Cognitive Sciences, 12*(5), 187–192.

Carnachan, E., & Parnell, J. (Writers). (2013). Restoring hope: An indigenous response to justice [Series episode]. In J. Parnell (Producer), *Pakip meka Aotearoa* [Documentary]. New Zealand: Notable Pictures.

Chen, Y.-W., Milstein, T., Anguiano, C., Sandoval, J., & Knudsen, L. (2012). Challenges and benefits of community-based participatory research for environmental justice: A case of collaboratively examining ecocultural struggles. *Environmental Communication, 6*(3), 403–421. doi:10.1080/17524032.2012.698291.

Christens, B., Collura, J., & Tahir, F. (2013). Critical hopefulness: A person-centered analysis of the intersection of cognitive and emotional empowerment. *American Journal of Community Psychology, 52*(1–2), 170–184. doi:10.1007/s10464-013-9586-2.

Chuck D, & Jah, Y. (1997). *Fight the power: Rap, race, and reality*. New York, NY: Delacorte Press.

Clayton, S. D., & Myers, G. (2009). *Conservation psychology: Understanding and promoting human care for nature*. Hoboken, NJ: Wiley-Blackwell.

Clayton, S. D., & Opotow, S. (2003). Justice and identity: Changing perspectives on what is fair. *Personality and Social Psychology Review, 7*(4), 298–310.

Cloitre, M., Stolbach, B. C., Herman, J. L., van der Kolk, B., Pynoos, R., Wang, J., & Petkova, E. (2009). A developmental approach to complex PTSD: Childhood and adult cumulative trauma as predictors of symptom complexity. *Journal of Trauma Stress, 22*(5), 399–408. doi:10.1002/jts.20444.

Cook, J. W. (1999). *Morality and cultural differences*. New York, NY: Oxford University Press.

Damon, W. (2009). *The path to purpose: How young people find their calling in life*. New York, NY: Free Press.

Decety, J., & Cowell, J. M. (2014). The complex relation between morality and empathy. *Trends in Cognitive Sciences, 18*(7), 337–339.

Decety, J., & Jackson, P. L. (2006). A social-neuroscience perspective on empathy. *CDIR: Current Directions in Psychological Science, 15*(2), 54–58.

Decety, J., & Michalska, K. J. (2010). Neurodevelopmental changes in the circuits underlying empathy and sympathy from childhood to adulthood. *Developmental Science, 13*(6), 886–899.

Drewnowski, A., & Specter, S. (2004). Poverty and obesity: The role of energy density and energy costs. *The American Journal of Clinical Nutrition, 79*(1), 6–16.

Dunbar-Ortiz, R. (2014). *An indigenous peoples' history of the United States*. Boston, MA: Beacon Press.

Duncan-Andrade, J. (2009). Note to educators: Hope required when growing roses in concrete. *Harvard Educational Review, 79*(2), 181–194.

Duncan-Andrade, J. (Producer). (2010, July 24). Note to educators: Hope is required when growing roses in concrete. *Askwith Forums* [Video]. Retrieved from http://www.gse.harvard.edu/news-impact/2010/02/note-to-educators-hope-required-when-growing-roses-in-concrete/.

Duncan-Andrade, J., & Morell, E. (2007). Critical pedagogy in an urban secondary classroom. In P. McLaren & J. L. Kincheloe (Eds.), *Critical pedagogy: Where are we now?* New York, NY: Peter Lang.

Edgcomb, G. S. (1993). *From swastika to Jim Crow: Refugee scholars at black colleges*. Malabar, FL: Krieger.

Edmonds, G., & Beardon, L. (Eds.). (2008). *Asperger Syndrome & social relationships: Adults speak out about Asperger Syndrome*. London, England: Jessica Kingsley.

Einolf, C. J. (2010). Does extensivity form part of the altruistic personality? An empirical test of Oliner and Oliner's theory. *Social Science Research, 39*(1), 142–151.

Einstein, A. (1946, June 23). The real problem is in the hearts of men: Professor Einstein says a new type of thinking is needed to meet the challenge of the atomic bomb. *New York Times Magazine*, SM4.

Eisenberg, C. (2011). *The wolf's tooth: Keystone predators, trophic cascades, and biodiversity*. Washington, DC: Island Press.

Environmental education framework: Creating an environment to educate about the environment. (2003). Nairobi, Kenya: United Nations Environment Programme.

Everett, D. L. (2008). *Don't sleep, there are snakes: Life and language in the Amazonian jungle*. New York, NY: Pantheon Books.

Eyler, J. (2002). Reflection: Linking service and learning—linking students and communities. *Journal of Social Issues, 58*(3), 517–534.

Florence, M. D. (2007). *Diet quality and academic performance in Nova Scotia grade five students*. Ottawa, Ontario, Canada: Library and Archives Canada = Bibliothèque et Archives Canada. Retrieved from http://worldcat.org /z-wcorg/ database.

Flyvbjerg, B. (2001). *Making social science matter: Why social inquiry fails and how it can succeed again*. Oxford, England: Cambridge University Press.

Franklin, R. G., Nelson, A. J., Baker, M., Beeney, J. E., Vescio, T. K., Lenz-Watson, A., & Adams, R. B. (2013). Neural responses to perceiving suffering in humans and animals. *Social Neuroscience, 8*(3), 217–227. doi:10.1080/17470919.2013.763852.

Freeman, M. (1994). The philosophical foundations of human rights. *Human Rights Quarterly, 163*, 491–514.

Freire, P. (1998). *Pedagogy of freedom: Ethics, democracy, and civic courage*. Lanham, MD: Rowman & Littlefield.

Freire, P. (2000). *Pedagogy of the oppressed* (30th anniversary ed.). M. B. Ramos (Trans.). New York, NY: Continuum. (Original work published 1970)

Freire, P. (2004). *Pedagogy of indignation*. Boulder, CO: Paradigm Publishers.

Freire, P., & Freire, A. M. A. (1992/1994). *Pedagogy of hope: Reliving "Pedagogy of the oppressed."* New York, NY: Continuum.

Fruehwald, E. (2010). A biological basis of rights. *Southern California Interdisciplinary Law Journal*, *19*(2), 195–236.

Garbarino, J. (1996). Children and poverty in America. *National Forum*, *76*(3), 28.

Garbarino, J. (2011). *The positive psychology of personal transformation: Leveraging resilience for life change*. New York, NY: Springer.

Garbarino, J., Bradshaw, C. P., & Vorrasi, J. A. (2002). Mitigating the effects of gun violence on children and youth. *The Future of Children*, *12*(2), 73.

Gesch, C. B., Hammond, S. M., Hampson, S. E., Eves, A., & Crowder, M. J. (2002). Influence of supplementary vitamins, minerals and essential fatty acids on the antisocial behaviour of young adult prisoners. Randomised, placebo-controlled trial. *The British Journal of Psychiatry*, *181*, 22–28.

Gibbs, J. C. (2014). *Moral development and reality: Beyond the theories of Kohlberg*. Retrieved from http://www.myilibrary.com?id=506937

Gini, G., Pozzoli, T., & Hauser, M. (2011). Bullies have enhanced moral competence to judge relative to victims, but lack moral compassion. *Personality and Individual Differences*, *50*(5), 603–608.

Ginott, H. G. (1972). *Teacher and child: A book for parents and teachers*. New York, NY: Macmillan.

Glasser, H. (2007). Minding the gap. In A. Wals (Ed.), *Social learning towards a sustainable world: principles, perspectives, and praxis* (pp. 35–61). Wageningen, the Netherlands: Wageningen Academic.

Grande, S. (2004). *Red pedagogy: Native American social and political thought*. Lanham, MD: Rowman & Littlefield.

Greenwald, A. G., Pratkanis, A. R., Leippe, M. R., & Baumgardner, M. H. (1986). Under what conditions does theory obstruct research progress? *Psychological Review*, *93*(2), 216–229.

Gruenewald, D. A. (2004). A Foucauldian analysis of environmental education: Toward the socioecological challenge of the Earth Charter. *Curriculum Inquiry*, *34*(1), 71–107.

Guest, K. J. (2014). *Cultural anthropology: A toolkit for a global age*. New York, NY: Norton.

Hahn, K. (1960, July 20). Address by Dr. Kurt Hahn at the annual meeting of the Outward Bound Trust. Retrieved from http://www.kurthahn.org/writings/obt1960.pdf.

Haidt, J. (2003). Elevation and the positive psychology of morality. *Flourishing: Positive Psychology and the Life Well-Lived*, *275*, 289.

Haidt, J. (2006). *The happiness hypothesis: Finding modern truth in ancient wisdom*. New York, NY: Basic Books.

Harré, N. (2011). *Psychology for a better world: Strategies to inspire sustainability*. Auckland, New Zealand: Department of Psychology, University of Auckland.

Hastings, P., Zahn-Waxler, C., & McShane, K. (2006). We are, by nature, moral creatures: Biological basis of concern for others. In M. Killen & J. Smetana (Eds.), *Handbook of moral development* (pp. 483–516). Mahwah, NJ: Lawrence Erlbaum.

Hayden, T. (2002). *The Zapatista reader*. New York, NY: Thunder's Mouth Press/Nation Books.

Heavy Head, R., & Blood, N. (Producers). (2007). *Blackfoot influence on Abraham Maslow*. [Video]. Retrieved from http://blackfootdigitallibrary.com/en/asset/blackfoot-influence-abraham-maslow,-presented-narcisse-blood-and-ryan-heavy-head-university-mo.

Hoffman, M. L. (1981). Is altruism part of human nature? *Journal of Personality and Social Psychology*, 40(1), 121–137. doi:10.1037/0022-3514.40.1.121.

Hoffman, M. L. (2000). *Empathy and moral development: Implications for caring and justice*. Cambridge: Cambridge University Press.

hooks, b. (2003). *Teaching community: A pedagogy of hope*. New York, NY: Routledge.

Horkheimer, M., & Adorno, T. W. (2002). *Dialectic of enlightenment: Philosophical fragments* N. G. Schmid (Ed.), E. Jephcott (Trans.). Stanford, CA: Stanford University Press. (Original work published 1944)

Illich, I. (2011). *Deschooling society*. London, England: Marion Boyars. (Original work published 1970)

Jackson, P. W. (1993). *The moral life of schools*. San Francisco, CA: Jossey-Bass.

Jacobson, K., & Silverbush, L. (Writers), & Colicchio, T. (Producer). (2012). *A place at the table* [Documentary]. United States: Magnolia Home Entertainment.

Johns, D. (2013). The war on nature—turning the tide: Lessons from other movements and conservation history. In M. Bekoff (Ed.), *Ignoring nature no more: The case for compassionate conservation* (pp. 237–255). Chicago, IL: University of Chicago Press.

Johnston, B. R. (1994). Concluding remarks. In B. R. Johnston (Ed.), *Who pays the price? The sociocultural context of environmental crisis* (pp. 233–235). Washington, DC: Island Press.

Jones, A. P., Happe, F. G. E., Gilbert, F., Burnett, S., Viding, E., Jones, A. P.,...Viding, E. (2010). Feeling, caring, knowing: Different types of empathy deficit in boys with psychopathic tendencies and autism spectrum disorder. *Journal of Child Psychology and Psychiatry*, *51*(11), 1188–1197. doi:10.1111/j.1469-7610.2010.02280.x.

Kahn, P. H. (1999). *The human relationship with nature development and culture*. Cambridge, MA: MIT Press.

Kahn, P. H., Friedman, B., Gill, B., Hagman, J., Severson, R. L., Freier, N. G.,...Stolyar, A. (2008). A plasma display window?—the shifting baseline problem in a technologically mediated natural world. *Journal of Environmental Psychology*, 28(2), 192–199.

Kant, I. (1963). *Lectures on ethics*. New York, NY: Harper & Row.

Kaplan, S., & Berman, M. G. (2010). Directed attention as a common resource for executive functioning and self-regulation. *Perspectives on Psychological Science*, 5(1), 43–57.

KRS-One. (2009). *The gospel of hip hop: First instrument*. Brooklyn, NY: powerHouse Books: I Am Hip Hop.

Krzesni, R. (1997). Interview 32648. In *Visual history archive*. Los Angeles, CA: USC Shoah Foundation Institute for Visual History and Education.

Kudryavtsev, A., Stedman, R. C., & Krasny, M. E. (2011). Sense of place in environmental education. *Environmental Education Research, 18*(2), 229–250. doi:10.1080/13504622.2011.609615.

Laland, K. N., & Brown, G. R. (2011). *Sense and nonsense: Evolutionary perspectives on human behaviour* (2nd ed.). Oxford, England: Oxford University Press.

Lazarus, E. (2002). The new colossus. *Emma Lazarus: Selected poems and other writings*. G. Eiselein (Ed.). New York, NY: Broadview Press. (Original work published 1903)

Loeber, A., van Mierlo, B., Grin, J., & Leeuwis, C. (2007). The practical value of theory. In A. Wals (Ed.), *Social learning towards a sustainable world: Principles, perspectives, and praxis* (pp. 83–97). Wageningen, the Netherlands: Wageningen Academic.

Lorde, A. (1984). *Sister outsider: Essays and speeches*. Trumansburg, NY: Crossing Press.

Martens, W. H. J. (2005). Multidimensional model of trauma and correlated antisocial personality disorder. *Journal of Loss & Trauma, 10*(2), 115–129. doi:10.1080/15325020590908821.

Marx, K., & Engels, F. (2010). *Manifesto of the Communist Party*. (Original work published 1847) Retrieved from http://www.marxists.org/archive/marx/works/1848/communist-manifesto/ marxists.org. (n.d.).

Marx, K., & Guesde, J. (1880). The programme of the Parti Ouvrier. Retrieved from http://www.marxists.org/archive/marx/works/1880/05/parti-ouvrier.htm.

Maslow, A. H. (1943). A theory of human motivation. *Psychological review, 50*(4), 370–396. doi:10.1037/h0054346.

Massachusetts General Hospital. (2006). 11 vitamins for a healthy brain: To stay sharp, your brain needs the right nutrients. Here's what you need to know about these essential vitamins. *Mind, Mood & Memory, 2*(10), 1–2.

Mays, V. M., Cochran, S. D., & Barnes, N. W. (2007). Race, race-based discrimination, and health outcomes among African Americans. *Annual Review of Psychology, 58*, 201–225.

McWorter, L. (2010). Enemy of the species. In C. Mortimer-Sandilands & B. Erickson (Eds.), *Queer ecologies: Sex, nature, politics, desire* (pp. 73–101). Bloomington: Indiana University Press.

Moore, K. D. (2005). The truth of the barnacles. *Environmental Ethics, 27*(3), 265–277.

Morton, S. G. (1839). *Crania Americana: Or a comparative view of the skulls of various aboriginal nations of North and South America*. Philadelphia, PA: Simpson, Marshall and Co.

Moya, P. M. L. (2003). *Reclaiming identity: Realist theory and the predicament of postmodernism*. Berkeley: University of California Press.

Myers, O. E., Jr., Saunders, C. D., & Garrett, E. (2003). What do children think animals need? Aesthetic and psycho-social conceptions. *Environmental Education Research, 9*(3), 305–325.

Myers, O. E., Jr., Saunders, C. D., & Garrett, E. (2004). What do children think animals need? Developmental trends. *Environmental Education Research, 10*(4), 545–562.

Nader, K. (2011). Trauma in children and adolescents: Issues related to age and complex traumatic reactions. *Journal of Child & Adolescent Trauma, 4*(3), 161–180. doi:10.1080/19361521.2011.597373.

Nave, C. S., Sherman, R. A., Funder, D. C., Hampson, S. E., & Goldberg, L. R. (2010). On the contextual independence of personality: Teachers' assessments predict directly observed behavior after four decades. *Social Psychological and Personality Science, 1*(4), 327–334. doi:10.1177/1948550610370717.

Nietlisbach, G., Maercker, A., Rössler, W., Haker, H., Nietlisbach, G., Maercker, A.,...Haker, H. (2010). Are empathic abilities impaired in posttraumatic stress disorder? *Psychological Reports, 106*(3), 832.

Nygreen, K., Ah Kwon, S., & Sánchez, P. (2006). Urban youth building community: Social change and participatory research in schools, homes, and community-based organizations. *Journal of Community Practice, 14*(1–2), 107–123.

Ogawa, M. (1995). Science education in a multiscience perspective. *Science Education, 79*(5), 583–593.

Oliner, S. P., & Oliner, P. M. (1988). *The altruistic personality: Rescuers of Jews in Nazi Europe.* New York, NY, and London, England: Free Press and Collier Macmillan.

Opotow, S. (1990). Moral exclusion and injustice: An introduction. *Journal of Social Issues, 46*(1), 1–20.

Opotow, S., & Weiss, L. (2000). New ways of thinking about environmentalism: Denial and the process of moral exclusion in environmental conflict. *Journal of Social Issues,* 56(3), 475–490.

Palmer, P. J., & Zajonc, A. (2010). *The heart of higher education: A call to renewal; Transforming the academy through collegial conversations.* San Francisco, CA: Jossey-Bass.

Pemment, J. (2013). Psychopathy versus sociopathy: Why the distinction has become crucial. *Aggression and Violent Behavior, 18*(5), 458–461.

Pengelly, M. (2014, May 24). Seven dead including gunman in "mass murder" California shooting. *The Guardian.* Retrieved from http://www.theguardian.com/world/2014/may/24/drive-by-shooting-near-uc-santa-barbara-leaves-seven-dead.

Plous, S. (2003). Is there such a thing as prejudice towards animals? In S. Plous (Ed.), *Understanding prejudice and discrimination* (pp. 509–528). Boston, MA: McGraw-Hill.

Pomeroy, D. (1994). Science education and cultural diversity: Mapping the field. *Studies in Science Education Studies in Science Education, 24*(1), 49–73.

Potter, R. A. (1995). *Spectacular vernaculars: Hip-hop and the politics of postmodernism.* Albany, NY: SUNY Press.

Powers, J. L., & Tiffany, J. S. (2006). Engaging youth in participatory research and evaluation. *Journal of Public Health Management and Practice, 12,* S79–S87.

Rahman, A. (n.d.). Reverse racism. Retrieved from http://poetry.rapgenius.com/Aamer-rahman-reverse-racism-annotated.

Rodger, E. (Producer). (2014). *Life is so unfair because girls dont want me* [Video]. Retrieved from https://www.youtube.com/watch?v=7KP62TE1prs.

Rodgers, Y. V., & Milewska, M. (2007). Food assistance through the school system: Evaluation of the Food for Kids program. *Journal of Children & Poverty, 13*(1), 75–95.

Rogers, P. J. (2001). A healthy body, a healthy mind: Long-term impact of diet on mood and cognitive function. *Proceedings of the Nutrition Society,* 60(1), 135–143.

Rosenthal, R. (1987). "Pygmalion" effects: Existence, magnitude, and social importance. *Educational Researcher, 16*(9), 37–41. doi:10.2307/1175728.

Rosenthal, R., & Jacobson, L. (1968). Pygmalion in the classroom. *The Urban Review, 3*(1), 16–20. doi:10.1007/BF02322211.

Sacks, E. L. (1952). Intelligence scores as a function of experimentally established social relationships between child and examiner. *The Journal of Abnormal and Social Psychology, 47*(Suppl. 2,), 354–358. doi:10.1037/h0060151.

Sayre, K. (1991). An alternative view of environmental ethics. *Environmental Ethics, 13*(3), 195–213.

Schapiro, R. A. (2005). The consequences of human rights foundationalism. *Emory Law Journal, 54*, 171–185.

Schoenthaler, S., Amos, S., Doraz, W., Kelly, M.-A., Muedeking, G., & Wakefield, J. (1997). The effect of randomized vitamin-mineral supplementation on violent and non-violent antisocial behavior among incarcerated juveniles. *Journal of Nutritional & Environmental Medicine, 7*(4), 343–352.

Schultz, P. W. (2000). New environmental theories: Empathizing with nature; The effects of perspective taking on concern for environmental issues. *Journal of Social Issues, 56*(3), 391–406. doi:10.1111/0022-4537.00174.

Severson, R. L., & Kahn, P. H. (2010). In the orchard: Farm worker children's moral and environmental reasoning. *APPDEV: Journal of Applied Developmental Psychology, 31*(3), 249–256.

Simpson, S., & Hsih, H. (2003). *The leader who is hardly known: Self-less teaching from the Chinese tradition*. Oklahoma City, OK: Wood 'N' Barnes.

Singer, P. (1975). *Animal liberation: A new ethics for our treatment of animals*. New York, NY: New York Review.

Smiley, T., & Ritz, D. (2014). *Death of a King: The real story of Dr. Martin Luther King Jr.'s final year*. Boston, MA: Little, Brown.

Sobel, D. (2008). *Childhood and nature: Design principles for educators*. Portland, ME: Stenhouse.

Spring, J. H. (2014). *American education* (16th ed.). New York, NY: McGraw-Hill.

Steffen, W., Grinevald, J., Crutzen, P., & McNeill, J. (2011). The Anthropocene: Conceptual and historical perspectives. *Philosophical Transactions of the Royal Society of London A: Mathematical, Physical and Engineering Sciences*, 369(1938), 842–867.

The Tbilisi Declaration. (1978). *Ambio, 7*(1). doi:10.2307/4312338.

Thomson, A. L., & Siegel, J. T. (2013). A moral act, elevation, and prosocial behavior: Moderators of morality. *The Journal of Positive Psychology*, 8(1), 50–64. doi:10.1080/17439760.2012.754926.

Tilbury, D. (2007). Learning based change for sustainability. In A. Wals (Ed.), *Social learning towards a sustainable world: Principles, perspectives, and praxis* (pp. 117–131). Wageningen, the Netherlands: Wageningen Academic.

Tilton, B. (2010). *Wilderness first responder: How to recognize, treat, and prevent emergencies in the backcountry*. Guilford, CT: FalconGuides.

TNL. (2003). *Recuerdo mis raíces y vivo mis tradiciones / Remembering my roots and living my traditions*. New York, NY: Scholastic.

Turiel, E. (2005). Resistance and subversion in everyday life. In L. P. Nucci (Ed.), *Conflict, contradiction, and contrarian elements in moral development and education* (pp. 3–20). Mahwah, NJ: Lawrence Erlbaum.

Ulrich, R. S. (1993). Biophilia, biophobia, and natural landscapes. In S. R. Kellert & E. O. Wilson (Eds.), *The biophilia hypothesis* (pp. 73–137). Washington, DC: Island Press.

UNEP. (1972). *Declaration of the United Nations Conference on the Human Environment.* Paper presented at the United Nations Conference on the Human Environment, Stockholm, Sweden. Retrieved from http://www.unep.org/Documents.Multilingual/Default.asp?documentid=97&articleid=1503.

UNEP. (1975). *The Belgrade Charter: A framework for environmental education.* Paper presented at the International Workshop on Environmental Education, Belgrade, Serbia. Retrieved from http://unesdoc.unesco.org/images/0001/000177/017772eb.pdf.

UNESCO. (1978). *Intergovernmental Conference on Environmental Education: Final report.* Paris, France: Author.

United Nations, General Assembly. (1948). *Universal declaration of human rights.* Retrieved from http://www.un.org/en/documents/udhr/

Visconti, L. (2009). Why the 'B' in 'Black' is capitalized at Diversityinc. Retrieved from http://www.diversityinc.com/ask-the-white-guy/why-the-b-in-black-is-capitalized-at-diversity-inc/.

von Thater-Braan, R., & Nelson, M. (Producers). (2012). *Grandfather, how do I learn? Exploring the foundations of diversity* [Video]. Retrieved from http://vimeo.com/71449994.

Wals, A. E., Beringer, A., & Stapp, W. (1990). Education in action: A community problem-solving program for schools. *Journal of Environmental Education, 21*(4), 13–19.

Weaver-Hightower, M. B. (2011). Why education researchers should take school food seriously. *Educational Researcher,* 40(1), 15–21. doi:10.3102/0013189X10397043.

Weiner, E. (2007). Critical pedagogy and the crisis of imagination. In P. McLaren & J. L. Kincheloe (Eds.), *Critical pedagogy: Where are we now?* New York, NY: Peter Lang.

Woodhouse, A., & Mark, A. (2012). The relationship of food and academic performance: A preliminary examination of the factors of nutritional neuroscience, malnutrition, and diet adequacy. *Christian Perspectives in Education,* 5(1), 1.

Yellow Bird, M., & Wilson, A. C. (2012). *For Indigenous minds only: A decolonization handbook.* Santa Fe, NM: School for Advanced Research Press.

INDEX

D

E

F

G

H

I

J

K

L

M

N

O

P

COUNTERPOINTS

Studies in the Postmodern Theory of Education

General Editor
Shirley R. Steinberg

Counterpoints publishes the most compelling and imaginative books being written in education today. Grounded on the theoretical advances in criticalism, feminism, and postmodernism in the last two decades of the twentieth century, Counterpoints engages the meaning of these innovations in various forms of educational expression. Committed to the proposition that theoretical literature should be accessible to a variety of audiences, the series insists that its authors avoid esoteric and jargonistic languages that transform educational scholarship into an elite discourse for the initiated. Scholarly work matters only to the degree it affects consciousness and practice at multiple sites. Counterpoints' editorial policy is based on these principles and the ability of scholars to break new ground, to open new conversations, to go where educators have never gone before.

For additional information about this series or for the submission of manuscripts, please contact:

Shirley R. Steinberg
c/o Peter Lang Publishing, Inc.
29 Broadway, 18th floor
New York, New York 10006

To order other books in this series, please contact our Customer Service Department:

(800) 770-LANG (within the U.S.)
(212) 647-7706 (outside the U.S.)
(212) 647-7707 FAX

Or browse online by series:

www.peterlang.com

www.ingramcontent.com/pod-product-compliance
Ingram Content Group UK Ltd.
Pitfield, Milton Keynes, MK11 3LW, UK
UKHW021851100926
8556IPUK00020B/1173

9 781433 131547